D1321125

MOTORWAYS

Motorways

JAMES DRAKE

C.B.E., BSc., F.I.C.E., F.I.Mun.E., P.P.Inst.H.E.

In association with

H. L. YEADON

B.Sc.Tech., M.I.C.E., M.I.Mun.E., A.M.Inst.H.E.

and

D. I. EVANS

M.S.E., B.Sc.Tech., F.I.C.E., F.I.Mun.E.

FABER AND FABER

24 Russell Square, London

First published in 1969
by Faber and Faber Limited
24 Russell Square London WC1
Printed in Great Britain by
Western Printing Services Ltd Bristol

SBN 571 09067 2

Contents

		Page
PREFACE		17
CHAPTER 1	WHY MOTORWAYS?	
	Main features	21
	Their purpose	22
	The benefits	26
CHAPTER 2	THE HISTORY OF MOTORWAYS	
	Italy leads the way	27
	The German State Motor Roads	28
	Development in the U.S.A.	30
	Britain's early interest	34
CHAPTER 3	POST-WAR DEVELOPMENT	
	The world awakens	42
	Progress in Britain	43
	Germany	49
	Italy	50
	France	51
	Belgium	52
	Holland	52
	Sweden, Denmark and Norway	53
	Austria	54

Switzerland *page* 54
South Africa 54
Japan 56
Canada 56
U.S.A.–The National System of Interstate and Defense Highways 57
U.S.A.–Toll Roads 61

CHAPTER 4 JUSTIFICATION AND PROCEDURE FOR A MOTORWAY IN BRITAIN
Introduction 67
Guiding principles of procedure 67
Co-ordination of steps in procedure 68
Justifying the need for a motorway 70
Will the motorway attract sufficient traffic, provide a convenient route and relieve congested roads? 72
What are the conditions on the existing road system? 73
Examples of economic assessments 75
Selecting the route 78
Fixing the line of the motorway 79
Alteration of side roads, footpaths and private accesses 81
Purchase of land 82
Design and contract documents 84
The final steps 86
Summary 87

CHAPTER 5 LAND ACQUISITION AND PUBLIC RELATIONS
The approach to the people affected 88
Land plans and schedules 89
Negotiations for land acquisition 91

The number of legal interests affected *page* 92
Compulsory purchase 93
The broad principles of compensation 94
During construction 95
Acquisition for county motorways 95

CHAPTER 6 MOTORWAY DESIGN
Introduction 96
Traffic 96
Carriageway capacity 97
Standards of alignment 97
Cross-section standards 101
Blending the motorway into the landscape 102
Survey of the route 104
Soil survey and its influence on design 105
Types of pavement 108
Drainage 112

CHAPTER 7 INTERCHANGES
Introduction 115
The guiding principles of design 115
Spacing of interchanges 116
Types of interchange 117
Two-level interchanges 117
Three-level interchanges 122
Four-level interchanges 124
The procedure of design 125
Some examples 126

CHAPTER 8 BRIDGES
The principal features 135
Motorway bridges in Lancashire 144

CHAPTER 9 CONTRACT PROCEDURE
Conditions of Contract 149

Drawings *page* 154
Specification 154
Bill of Quantities 166
Future trends in contract procedure 167

CHAPTER 10 THE CONSTRUCTION OF THE MOTORWAY
Some aspects of construction 169
Examples of completed motorways 173

CHAPTER 11 MAINTENANCE AND OPERATION OF MOTORWAYS
Maintenance 183
Policing and emergency services 187
Aids to movement of traffic 192
Service areas 194
Speed limits on motorways 195

CHAPTER 12 URBAN MOTORWAYS
General 199
Design standards 199
Interchanges 200
Tidal and peak hour flow of traffic 202
Elevated or sunken? 203
Structures 204
Examples of urban motorways 205

POSTSCRIPT 208

INDEX 211

Figures

		Page
1	Types of accidents with turning or crossing traffic	23
2	Motorways proposed by Institution of Highway Engineers, 1936	37
3	National Plan for Motorways, County Surveyors' Society, 1938	38
4	Ministry of Transport Road Programme, 1946	44
5	Motorways in Britain, 1968	48
6	Motorways in Europe	55
7	National System of Interstate and Defense Highways–U.S.A.	58
8	Network analysis for a typical motorway project	69
9	Cost per mile for average vehicle varying with speed	74
10	South Lancashire Motorway	75
11	Preston Northerly By-pass	77
12	Part of Draft Scheme Plan for Line of Motorway. Lancashire-Yorkshire Motorway, M.62	80
13	Draft Scheme Plan for connecting roads at interchange between M.62 and Bury New Road, A.56	80
14	Draft Order Plan for side road alterations, etc., near Bury New Road, A.56 for M.62	83
15	Draft Order Plan for trunk road alterations to Bury New Road A.56 for M.62	84
16	Part of contract drawing for Lancashire-Yorkshire Motorway	85
17	Snowhill Lane bridge on M.6. Perspective sketch	85
18	Land reference plan	90
19	Land reference schedule	90

FIGURES

20 Plot (or interest) plan *page* 91
21 Current Ministry of Transport motorway cross-section for rural areas 101
22 Two-level interchanges 118
23 Three-level interchanges 123
24 Interchange between Lancashire-Yorkshire Motorway, M.62, and North-East Lancashire Motorway 123
25 Four-level interchanges 124
26 Worsley braided interchange between M.62, M.61, A.580 and A.666 129
27 Worsley braided interchange. Traffic figures–vehicles per day in 1979 130
28 Eccles interchange between South Lancashire Motorway and Stretford-Eccles By-pass, M.62 132
29 Croft interchange between South Lancashire Motorway and M.6 133
30 Tarbock interchange between South Lancashire Motorway and Liverpool Outer Ring Road 134
31 Motorways in Lancashire 174
32 Typical cross-section for Thelwall to Preston length of M.6 178
33 Typical cross-section for the Preston-Lancaster Motorway, M.6. 181
34 'Scissors' crossing of slip roads 201
35 Slip roads joining motorway at right angles 201
36 Collector-distributor roads 202

Tables

		Page
1	Number of injury-accidents at each type of junction	23
2	Accident rates on completed motorways in Lancashire	25
3	Saving in accidents due to constructing the 13½ miles of M.6. between Preston and Lancaster By-passes	25
4	Mileage of motorways completed each year in England and Wales	47
5	Number of drawings prepared for M.6 in Lancashire	86
6	Rural motorway capacities	97
7	Casualties on motorways in Lancashire County Police area	198

Plates

Between pages 96 and 97.

1 Grade-separated roundabout on the M.6 at A.6, Bamber Bridge, Preston, 29/M.6
2 Broughton interchange, 32/M.6
3 Barton high-level bridge
4 Thelwall bridge
5 Lune bridge
6 Gathurst viaduct
7 Lodge Lane bridge
8 Samlesbury bridge
9 Fylde junction higher bridge
10 Snowhill Lane bridge
11 Parkhead Lane bridge
12 Mount South bridge
13 Jeps Lane bridge
14 Kenlis Arms bridge
15 Shaft footbridge
16 Lydiate Farm footbridge
17 Example of old mine gallery encountered during excavation
18 Example of old mine shaft encountered during excavation
19 Maintenance compound on M.6
20 Gritting vehicle
21 Mowing machine developed for use on steep slopes
22 Charnock Richard service area on M.6
23 Forton service area on M.6

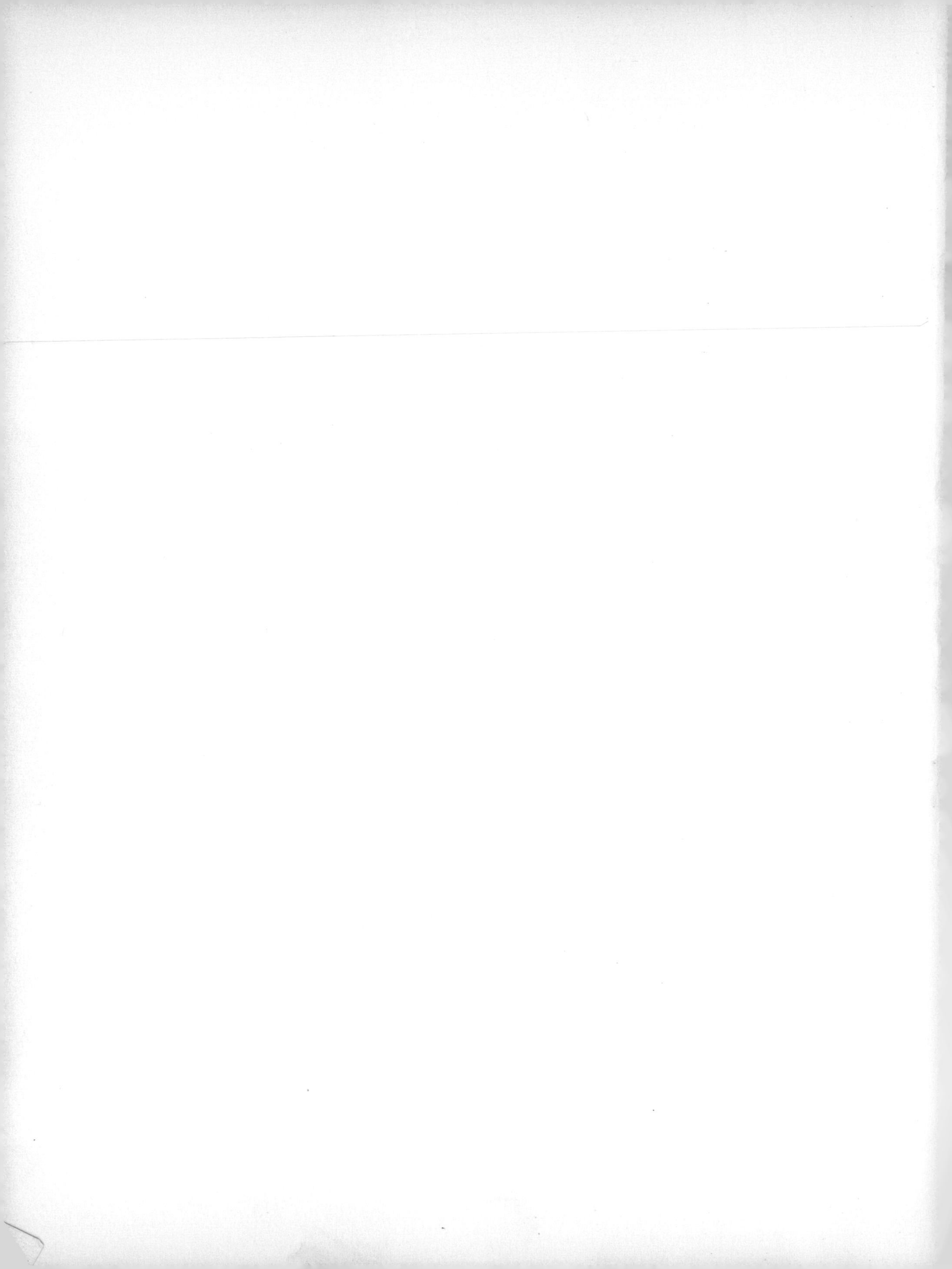

Preface

My interest in motorways started in 1938 when, as the newly appointed Surveyor to the County Borough of Blackpool, I was a member of a party of engineers who visited the autobahnen at that time under construction in Germany.

So convinced was I that they were the only type of road suitable for carrying the high volumes of traffic on Great Britain's major routes of the future, that I recommended my Committee in 1939 to build the proposed Blackpool Ring Road as a motorway. It is still planned as a motorway and is still 'proposed'.

In 1945, when I was appointed Surveyor and Bridgemaster to my native County of Lancashire, I was excited by the thought that, at an early date, I might be designing a motorway to replace the trunk road running north-south through the County. My predecessor in office, Mr. Peter Schofield, had recommended this solution to his Committee in 1938; and it was, in fact, incorporated in the National Motorway Plan recommended to the Ministry of Transport by the County Surveyors' Society. In 1944 the Ministry of War Transport had considered that there was justification for examining the possibilities of making the proposed north-south route through Lancashire a motorway.

I set about furthering this scheme as soon as I could after taking up my new appointment, little realising that it would be another 13 years before the first section, viz. the Preston By-pass length of M.6, would be completed. Incidentally, it was the first motorway to be completed in Great Britain and provision for its construction (to be started in 1956) had been made in the expanded road programme, which Mr. A. T. Lennox-Boyd as Minister of Transport and Civil Aviation announced to the House of Commons on the 8th December, 1953.

It was on this scheme that one of my Associate Authors, Mr. H. L. Yeadon, who has been with me since 1948, had his first experience of motorway construction as an Assistant Resident Engineer. After further motorway experience he was appointed, in 1962, Resident Engineer on the construction of the M.6 between Preston and Lancaster; and he is now one of my three Chief Assistants, specialising in construction.

My interest in motorways has taken me abroad on many occasions, including four visits to the U.S.A. and Mr. Yeadon accompanied me on two of these four visits. In May 1965 I had the honour of leading a party of 33 members of the County Surveyors' Society to the United States and Canada to make a most extensive study of their highway work.

My other Associate Author, Mr. D. I. Evans, has also been to the U.S.A. Under the auspices of the English-Speaking Union, he was appointed Doty Fellow and Assistant in Instruction at Princeton University, New Jersey, for the academic year 1953–54. This was followed by experience of American road construction with a firm of Consulting Engineers. Mr. Evans joined my staff in 1959 and is at present a Senior Engineer specialising in the design of motorways.

During my period as County Surveyor and Bridgemaster of Lancashire, my Department was responsible for designing and supervising the construction of 67 miles of motorways–61 miles as agent for the Ministry of Transport and 6 miles on behalf of the Lancashire County Council. A further 50 miles were either under construction or at the design stage when I was seconded to the Ministry of Transport in February 1967. Although some of these motorways pass through built-up areas, they can be looked upon as inter-town motorways and this book is mainly concerned with motorways of this kind. Urban motorways could well provide the subject for a book of their own, but their characteristics are described in Chapter 12.

For the sake of easier reading, I have generally divided the descriptions of motorways in various countries into those built before 1945 and those built since. Lack of space has prevented the inclusion of many countries. I have omitted those which I thought might be better left to writers with a more intimate knowledge of them. For similar reasons in dealing with this Country I have concentrated on motorway work with which I have personally been associated and, where appropriate, I have put forward my own views.

Highway transportation plays an important role in the economic and social life of any nation. It is, of course, only one form of transportation–railways, airways, waterways and pipelines have their contributions to make. However, the wider field of transport systems and their integration is considered to be outside the scope of this book. What is certain is that our system in Britain has become more and more overloaded, and that every effort must be made to provide a motorway system which is better able to cope with the present and future motor transport.

Civil engineers are traditionally regarded as a somewhat reticent body of men, pursuing their task of directing the sources of power in nature for the use and convenience of man, often without explaining to the public the reasons behind their endeavours. There have been many articles about motorways in the technical press and papers on various aspects of the subject have been read to professional institutions, but there have been few publications which attempt to set out in plain language, with the minimum of statistics and engineering terminology, the general principles, philosophy and logic associated with motorways. It is my hope, therefore, that this book might help to fill a gap. It is intended for laymen as well as the civil engineering profession; and with this in mind we have given an explanation wherever appropriate of those technical terms which have had to be used.

I wish to place on record my appreciation for the help of the many friends who have assisted in providing information and advice.

In particular, I would like to express my thanks for the contributions made by Mr. W. J. H. Palfrey, O.B.E., Chief Constable of Lancashire, Mr. Arthur E. Bowles, C.B.E., G.I.F.E., the County's former Chief Fire Officer, Dr. S. C. Gawne, M.D., B.S., M.R.C.S., the County's Medical Officer of Health, Mr. W. Skellern, B.Sc.Tech., M.I.Mun.E., Dip.T.P., Mr. F. Lawton, A.A.I., and Mr. C. Kay, A.M.Inst.H.E.

To my brother-in-law, Mr. H. Crossley, I am very much indebted for his extremely useful editorial assistance and to Mr. Ernest Davies, Editor of *Traffic Engineering and Control*, for editing the original manuscript and reducing it to a manageable length.

JAMES DRAKE

Chapter 1

Why motorways?

Main features

The need for motorways under modern traffic conditions is obvious to every road engineer and many others, too. Unfortunately, in this Country, this is an opinon which has not always been universally upheld.

I well remember the occasion–at a very delicate stage in the negotiations for the start on the Preston By-pass section of the M.6–when articles appeared in the Press headlined 'Motorways will be Murderways'.

Fortunately a more enlightened view prevailed at the Ministry of Transport and there was no delay in starting the work.

What a tragedy it would have been if these articles had been taken seriously. It requires only the temporary closure of one of our motorways today to bring home the point most strongly.

The main characteristics of a motorway in Britain are:

1. The exclusion of pedestrians, pedal cycles, horsedrawn traffic, agricultural tractors, mopeds, motor cycles under 50 c.c.'s, and certain other classes of motor vehicles;
2. the complete separation of opposing traffic streams;
3. the elimination of all crossings at the same level;
4. the provision of specially designed interchanges at reasonably long intervals and the elimination of all other accesses;
5. moderate gradients and gentle curves with adequate sight distances;
6. hard shoulders on the nearside of each carriageway in order that vehicles may park completely off the carriageways in cases of emergency or mechanical failure;
7. telephones at one-mile intervals on each side of the motorway for use in emergency;
8. servicing and refreshment facilities provided initially at approximately 24-mile intervals; and

9. Police posts at approximately 24-mile intervals and highway maintenance depots at closer spacing.

Their purpose

The purpose of motorways is fairly obvious: they should provide safer travelling, they should reduce the time taken to make the journey, and they should carry more traffic.

A. Safer travelling

Since all road accidents involving personal injuries must be reported to the police, it is a simple matter to obtain records of all injuries, including fatal accidents, on any particular stretch of road.

If the volume of traffic is also known over this road section, then it is easy to calculate the accident rate in terms of vehicle-miles. The normal standard is the number of accidents per million vehicle miles and, for fatal accidents, where the proportion is naturally much smaller, per hundred million vehicle-miles.

The accident rate for all-purpose roads depends on a number of factors. First, the type of road. As one would expect, dual carriageways are safer than single carriageways because head-on crashes are reduced. The standard for British motorways provides for dual carriageways and, although the width of the central reserve is not as wide as many highway engineers would like, its provision reduces the number of head-on collisions compared with single carriageway all-purpose roads.

Surfacing, too, is vitally important. I believe it is true to say that Britain takes greater pains than most other countries to try to produce the best possible surface from the point of view of skid resistance. No effort is spared to provide these good surfaces but even the best have a very low coefficient of friction when wet, with speeds approaching 70 miles per hour. This is one of the reasons why, for many years now, I have supported this speed as the limit on British motorways.

Accidents involving pedestrians and pedal cyclists can account for an appreciable proportion of accidents on all-purpose roads, the rate generally increasing with the density of building development alongside. The simple fact that both pedestrians and pedal cyclists are banned from motorways completely cuts out these accidents, and the reduction in accidents is most marked where a motorway by-passes a built-up area.

Similarly, every entrance from adjacent property, be it from houses, shops, factories, schools–even from agricultural land–provides a potential accident spot. No such entrances are allowed on motorways.

When all the injury-accidents on 205 miles of trunk and classified roads in Lancashire were fully investigated in the year 1946–7, it was found that 45 per cent of the accidents had occurred at intersections.[1] The various types of accidents which can occur to turning or crossing traffic are shown in Figure 1.

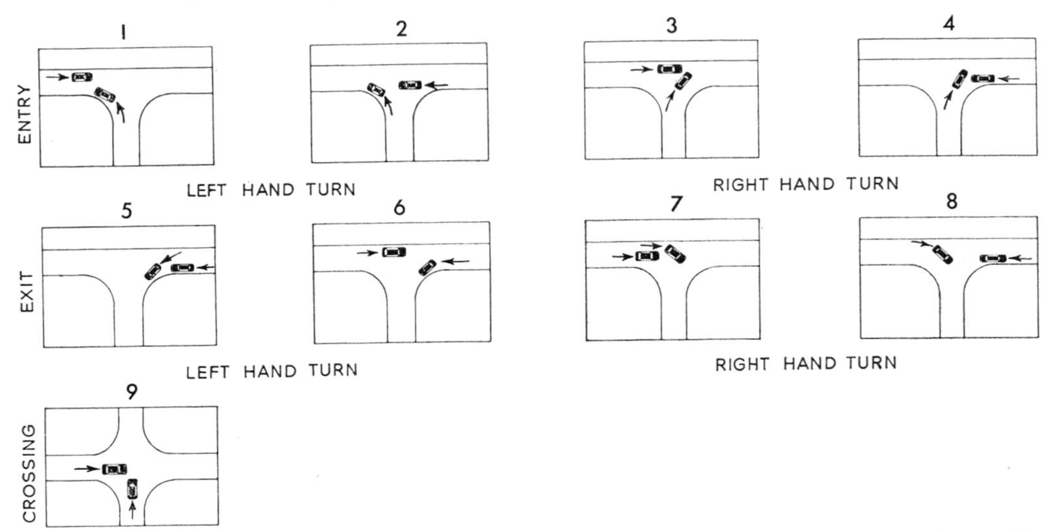

Fig. 1. Types of accidents with turning or crossing traffic

The investigation showed that the number of injury-accidents occurring at each type of junction were as given in Table 1.

Table 1
Number of injury-accidents at each type of junction

Type of accident	1	2	3	4	5	6	7	8	9	Total
Number	0	18	8	22	4	0	47	25	21	145

Had the traffic under survey been travelling on motorways, a large proportion of these intersection accidents would have been rendered impossible by the very nature of the motorway design.

If we list them, we arrive at a hypothetical reduction in accidents of 123, made up as follows:

Through central reserve of motorway (types 1 and 6)	Nil
No right-turns on to motorway (types 3 and 4)	30
No right-turns off motorway (types 7 and 8)	72
No crossing accidents (type 9)	21
Total	123

[1] Lancashire County Council, *Road Plan for Lancashire*, 1949.

To make the figures fair, however, the statistician would argue that on motorways the number of left-hand turns is double that on other types of roads and that we must, therefore, multiply the 22 accidents due to left-hand turns (types 2 and 5) by two. Thus the reduction in the number of intersection accidents, had the roads investigated been motorways, would have been from 145 to 44, that is to say, a net reduction of 101.

To assume that doubling the number of left-hand turns on motorways would double the number of accidents is, however, ignoring the difference in the way a left-turn is made to and from a motorway. In practice, the provision of acceleration and deceleration lanes at interchanges, the improvement of the sight line and the concentration of entering and leaving traffic to fewer junctions results in a reduction rather than an increase of left-turn accidents.

The 1946 investigation showed also that 20 per cent of the accidents on rural main roads would have been prevented if standing vehicles had not been allowed on the carriageway. A subsequent investigation by the Road Research Laboratory gave the figure as varying between 12 per cent on two-lane roads to 23 per cent on three-lane roads.

On motorways, accidents due to standing vehicles have been almost entirely eliminated by the provision of hard shoulders on to which all vehicles must move before coming to a halt, and then only in an emergency.

Lastly there is the question of curvature of the road. As the curvature increases so does the road hazard. The motorway's gentle curves and longer sight line are other positive factors towards the reduction in accidents.

Conclusion on safer travelling on motorways. After considering all these factors and studying the accident rates on foreign motorways, I made a forecast many years ago that the injury-accident rate on motorways would be approximately 0·40 per million vehicle-miles. The actual rate to date for the lengths of motorways opened at various times in Lancashire is shown in Table 2 and a comparison is made with the accident rate on the superseded all-purpose roads.

Table 3 shows the number of various types of injury-accidents which occurred on the all-purpose road A.6 during the twelve months before the length of M.6 was built alongside, as compared with the same type of accidents on this length of A.6 and M.6 for the 12 months immediately following the opening of the motorway. There was a net saving of 60 accidents despite a total increase in traffic of 32 per cent. In this case the length of A.6 by-passed traverses a more or less rural area and consequently has a lower accident rate.

The other lengths of motorway in Lancashire by-pass all-purpose roads which

Table 2

Accident rates on completed motorways in Lancashire

Description	Injury-accident rate per million vehicle-miles		Fatal accident rate per 100 million vehicle-miles	
	Motorway	Superseded route	Motorway	Superseded route
M.62 Stretford-Eccles	0·364	5·1	3·10	18
M.6 Preston By-pass	0·248	5·8	1·60	20
M.6 Lancaster By-pass	0·376	4·6	3.74	16
M.6 Preston-Warrington	0·251	3·56	3·24	14
M.6 Broughton-Hampson	0·258	1·45	*	9

* Only two fatal in the first two years of operation.

Table 3

Saving in accidents due to constructing the 13½ miles of M.6 between Preston and Lancaster By-passes

	12 months before opening	12 months after opening	
	on A.6	on A.6	on M.6
Fatal	8	4	0
Serious	29	15	4
Slight	106	41	19
Total	143	60	23

pass through more densely built-up areas. Table 2 shows that their accident rate is much higher, with the result that the saving in accidents through these lengths of motorway will also be far higher than in the motorway in Table 3.

B. **Reduction in journey time**

With regard to reduction in journey time, one must not overlook the fact that on typical all-purpose roads (with intersections on the level, frontage development often requiring a 30 or 40 miles per hour speed limit, the presence of pedestrians, cyclists and, on occasion, horse-drawn traffic) the accident rate results from a much slower traffic flow than that on the motorway, where not only cars but also heavy traffic, such as coaches and lorries without trailers, can travel up to the present general speed limit of 70 miles per hour. Thus motorways are not only safer, but safer at much higher speeds than all-purpose roads.

C. **Carrying more traffic**

Just as the strength of a chain is that of its weakest link, so the capacity of an all-purpose road is limited by its worst sections. Although bottlenecks in built-up areas may sometimes be critical, generally intersections at one level are the main cause of congestion. Interchanges on motorways are sited and designed to allow the road to achieve its maximum capacity. The prohibition of horse-drawn vehicles and pedal-cyclists also increases the carrying capacity of a motorway over that of an all-purpose road.

With vehicular traffic increasing at its present rate in our densely built-up country, it is particularly important to obtain the maximum capacity from our major roads.

The benefits

It is not just the traffic using the motorways that benefits by their construction: for example, a town situated in close proximity to a motorway benefits in two ways. Firstly, the town is relieved of its through traffic. Numerous towns now choked with traffic would have their normal life restored by the provision of motorways through or around them. Not only is traffic on the main roads delayed, but traffic has great difficulty in either entering or crossing the main roads.

Secondly, the proximity of a motorway makes a town far more attractive from an industrial point of view. Adequate road communications are always a prime consideration of an industrialist looking for a site for his new works; a convenient connection to the motorway system is a great advantage.

There is no doubt that the completion of M.6 through the centre of Lancashire has brought new life to the towns alongside it. A motorway connection from M.6 to Liverpool would, I believe, have a more stimulating effect on the industry of that city than any other scheme, including the very urgently required second tunnel under the Mersey.

In the same way, a motorway connection to a major urban area will allow the opening up for development of areas which otherwise would be too far distant in travelling time from major urban centres.

One further point which is not always appreciated is that motorways do not cut up urban areas in the same way as major all-purpose roads. They cross other roads either by bridging over or under them. Thus, in a built-up area, the motorway is generally either above or below ground level with frequent crossings under or over it. It is therefore possible to travel from one side of the motorway to the other in safety and a town is not virtually cut in two as is the case when a busy all-purpose road passes through it. There is no backing-up of traffic at interchanges on to the motorway and, in fact, the life of the town proceeds as if through-traffic were non-existent.

Chapter 2

The history of motorways

Italy leads the way

Italy was one of the first to construct roads exclusively for motor traffic. Principally responsible was Signor P. Puricelli, of Milan, contractor, financier and a distinguished engineer, who prepared a motor road plan in the 1920's which was supported by the Italian Touring Club and backed by Mussolini. In 1923, several state-subsidised companies were formed to construct and operate toll roads, which were to revert to state control after a defined period.

The autostrade were unlike present day motorways in that they consisted of only a single carriageway, sometimes with only three lanes and without segregation of traffic travelling in opposite directions as now. The roads were, however, a big step forward because to permit traffic to travel safely at high speed, along their entire length they were completely fenced off from adjoining land. Traffic could enter and leave only at controlled points and no other road or railway line crossed at the same level. Carriageway width varied from 26½ feet on some sections to 33 feet on others, with side verges of 3½ to 5 feet wide.

The first to be started was the Milan-Varese Autostrada in 1924. During the next ten years 330 miles were opened to traffic in eight main sections, the longest being the 80-mile section from Milan to Turin. With the exception of the 13-mile Naples-Pompeii length, the system was confined to the north of the country and by 1935 stretched from coast to coast.

For various reasons, mainly political, the rate of construction declined in the 1930's and the road building effort was transferred to the Italian colonies in North Africa where in 1937 a new road over 1,000 miles long across Tripoli was opened to traffic. Generals Rommel and Montgomery were both to find this extremely useful not many years later!

The German State Motor Roads

Ostensibly Hitler's purpose in ordering the building of a network of express highways was to alleviate Germany's high rate of unemployment in the 1930's, but a major reason was to provide a means of transporting motorised troops quickly across the country. Whatever his motives, the fact remains that Germany was the first country to plan a nation-wide system of highways, capable of meeting modern traffic needs, although express highways on a limited scale had already been constructed, notably in the U.S.A. and Italy.

Military considerations apart, Germany had special problems: many towns and cities were extremely old and, not only were their streets narrow, but they contained buildings of great historical and aesthetic value, so that extensive widening of city thoroughfares was out of the question. In addition the buildings were close together and the centres of population heavily concentrated. To attempt to solve the problem by by-passing the towns on the basis of the existing road system would merely have resulted in a series of by-pass roads, uneconomical in construction, and wasteful in mileage. Further, because the price of petrol was four times that in the U.S.A., it was even more important to have traffic flowing at a steady and reasonable speed which would, it was estimated, save 20 to 35 per cent in operating costs.

Hitherto German highway administration had been divided between 26 provincial and 600 independent township highway departments. Their failure to co-operate in the development of an effective modern highway system prompted Hitler's government to pass a law, in March 1934, delegating the responsibility to a single agency, a subsidiary of the German State Railways, called the 'Reichsautobahnen', with a capital of 50 million marks. The purpose of this relationship was to ensure that the experience of the Railways was fully utilised and also to promote their direct interest in road traffic.

The precise estimated cost of the scheme as planned is not known, but it is believed that it was in the region of 1,260 million marks, or say £63 million. Of the additional capital involved, 60 per cent was to be provided from saving in unemployment relief and the remainder from an anticipated increase in taxation revenue arising from improved national prosperity.

The 'Reichsautobahnen' was also given special rights over the trading incidental to the new roads, such as the operation of filling and repair stations, rest houses and the control of advertising. Some degree of preference was also given in respect of bus services.

The future revenue to meet expenditure was assured by the imposition of a levy of 5 pfgs. per litre of petrol sold at the official filling stations (equivalent to about

3d. per gallon) which was estimated to yield 600 million marks per annum (say £30 million) which was considered sufficient to provide for the costs of maintenance and the loan charges on the capital raised.

Dr. Fritz Todt was appointed Inspector General of German Road Construction, directly responsible to Hitler, with the task not only of locating and constructing the new 'Autobahnen' but also with the renewal and development of the old road network. He was given paramount authority in all matters of finance and the power of expropriation of land where it could not be acquired by negotiation. In the preparatory planning of the system he was to be assisted by a Committee representing interested bodies, including industrial undertakings.

The basic concept was for two principal routes running north and south and four east to west. The north-south routes comprised one in the west extending from Hamburg to the Swiss frontier at Basle and the other from Stettin on the Baltic, through Berlin and Munich to the Austrian border, near Salzburg. The initial programme provided for the construction of 4,500 miles of autobahn over a five-year period with the intention of increasing the mileage to 7,500 miles within ten years.

By the end of 1937 almost 1,000 miles had been opened to traffic with a further 1,100 miles under construction. The survey and design work for another 4,300 miles was also in progress.

Prior to the start of the War the mileage opened to traffic had increased to over 2,300 miles.

The general principles on which the new motor roads were planned sound familiar to modern ears, but in 1934 were quite revolutionary. The autobahnen were located to link together the main cities and industrial areas and a few of the largest cities were to be provided with ring roads as part of the system. In the majority of cases, however, the motor road was sited to pass within a few miles of the city and the connection was made by a specially constructed spur or by the normal road system.

The design standards adopted provided for the maximum flow of traffic at a minimum cost of operation with the greatest possible safety of movement. High traffic flow with low operating costs was to be achieved by laying a uniform road surface to high standards of accuracy and permanence; easy gradients and ample curves, with super-elevation where necessary; the highest attainable standards of visibility; and the provision of parking places and lay-bys off the highway.

In the cause of safety, all pedestrian, pedal cycle and animal traffic was to be excluded; opposing traffic streams completely separated; 'collision points' eliminated by construction of over- or under-bridges to prevent traffic streams crossing each other; traffic at road interchanges controlled by a clear and systematic layout for each class of interchange, and road signs were to be clear and simple. Lastly, there was to be no right of access from adjoining land.

In 1946 a group of British highway engineers, representing the Ministry of Transport, the Road Research Laboratory and the Cement and Concrete Association, visited Germany to inspect the autobahn system, to assess its post-war condition and to see what lessons could be learned.

It was found[1] that the layout and alignment was of a sufficiently high standard to suggest a degree of safety above that obtained on ordinary main roads in Britain. Several factors were considered to contribute to this. First, only motor vehicles were allowed to use the motor roads, pedestrians and all other vehicles being excluded; second, access was fully controlled and only permitted at the properly designed and comparatively infrequent interchanges; and third, the uniformity of design, especially in the width of carriageways, eliminated many driving hazards encountered elsewhere.

The advantage of providing a wide bankette at the side of the carriageway, where vehicles could stop in an emergency without causing obstruction, was noted. It was suggested, however, that a minimum width of 7 feet 6 inches was necessary with its surface in a contrasting colour to the carriageway.

Adverse comments were of a comparatively minor nature, and mainly concerned with faults in alignment. In the earlier construction very long straights had been employed, joined by curves only where necessary, and there was also evidence of lack of co-ordination between the horizontal alignment and the vertical profile which, it was considered, affected the appearance.

Whatever the faults, a bold attempt had been made and Germany had shown what could be done in establishing a modern highway system.

Development in the U.S.A.

The earliest design for a controlled access road for exclusive use of motor traffic was the Bronx River Parkway from Bronx Park in New York City into Westchester County, a distance of 15½ miles.

The scheme was first suggested in 1914 when the number of cars travelling out of the City into the quieter residential districts at weekends and holiday times was causing increasing concern. Construction was delayed by the outbreak of World War I so the Parkway was not completed until 1925.

Although the provision of dual carriageways was considered, the road was even-

[1] Department of Scientific and Industrial Research, Road Research Laboratory, *German Motor Roads 1946*. London, H.M.S.O. 1948.

tually built with four lanes, three of which were used on occasion to carry the heavy volume of traffic returning to the City after a day in the country. This is probably one of the earliest examples of applying the principle of tidal traffic flow now widely used in many countries. The road was designed for a speed of 35 miles per hour and lorries were completely excluded.

The creation of this valuable traffic facility had the secondary effect of restoring the appearance of the bank of the Bronx River, which had deteriorated over the years and become a dumping ground for nearby communities. Thus an area which had formerly contained some of the lowest priced land in the County, now swept clean of roadside development and advertising signs, increased in value so quickly that the cost of constructing the road was soon balanced by the increase in land taxes collected. There is a lesson here from which we have much to learn today, forty years later.

The immense success of the Parkway was followed by similar projects throughout Westchester County, and in the decade prior to 1933 over $80 million was spent on parkways there.

At the same time the Long Island State Park Commission, under its Chairman, Robert Moses, was developing another system on Long Island. Elsewhere in the country, notably in the Washington area and in Connecticut, similar systems were developing but not on as great a scale. Many of these roads had four lanes. Thus traffic in each direction was provided with a slow lane and a fast overtaking lane.

Although accident experience on four-lane roads was markedly better than those with two or three lanes, the need to separate opposing traffic streams soon became apparent. It was found that on such high speed roads the strain on drivers in the fast lane was excessive, particularly at night, when headlight glare from opposing vehicles had to be contended with. Under these conditions accidents were inevitable and usually involved several vehicles.

As an expedient on roads where the overall width was restricted, barriers were erected along the centre line, but where possible a median or central reserve was constructed segregating the two carriageways.

It is not precisely known when the first divided highway was built. There must, of course, be many instances where railways, tramways or waterways formed a division between carriageways. Probably the first road designed with a wide median was the Dupont Highway in Delaware, built in the 1920's, and even though it did not have control of access, it demonstrated the higher degree of safety obtainable. In 1934 California built its first rural divided highway, incorporating a median strip 20 feet wide. In the same year the Maryland Road Commission let contracts for a road with a 30 feet median, although some favoured a width of 50 feet. It was not until 1938, however, that California became probably the first State to build to this standard.

MOTORWAYS

In 1934 Mr. Moses, then Construction Co-ordinator for New York City, began the task of co-ordinating the work of design and construction of a system of parkways and expressways for the whole of the Metropolitan Area.

The aim of the system was to allow for rapid traffic movement around the entire circumference and through the interior of New York City. Connections were to be provided from the City to the bordering communities of New Jersey to the west, Westchester County to the north, and Long Island to the east.

Before World War II, 164 miles of four-lane and six-lane highway had been constructed in the Metropolitan Area. All one-level crossings were eliminated, the roads being carried over or under the motorways by bridges. Altogether 280 such grade separations were built.

The parkways were reserved for passenger cars only and, in this respect, differed from expressways and freeways. Many were provided essentially for commuters. Service roads, usually without grade separation, were provided alongside the parkways for commercial and local traffic.

Considerable effort was made to give the parkways a pleasing appearance. The width of the strip of acquired land, or 'right of way', was generally of about 300 feet but where it was desired to provide special treatment, the width was increased to 400 feet, and in heavily built-up areas the width was reduced to 200 feet. Landscaping was undertaken throughout, the median and side verges were laid out as lawns, and shrubs and trees were planted elswehere in profusion.

When I first visited the U.S.A. in 1960 and met Mr. Moses, he observed that in laying out the parkway programme in the 1930's, there were no illusions as to the tremendous task ahead. The need for modern arteries of travel, including recreational facilities, had become a pressing vital factor in the orderly development of the whole Metropolitan Area. To have delayed would have been a major public blunder. Intolerable conditions would have developed had the arterial parkway programme not been advanced to meet the traffic problem created by the sharp increase in motor car production and travel.

Because the years between the Wars were depression years making normal methods of financing difficult, after a break of nearly a century, there was a move towards the construction of toll roads or turnpikes. Turnpikes are a relic of earlier days, when a long staff attached to a wooden barrier was set across the road. This was the actual turnpike which was opened only when the traveller had paid the required toll. The term is still extensively used today in the naming of many of America's major toll roads.

Where the need was most pressing for the provision of roads specifically designed for high density, high speed traffic, toll roads were built. To provide for the collection of tolls, access had to be completely controlled, and this was immediately recognised as an enormous asset. Later, when the need for tolls was not so pressing and freeways were financed from taxes, many argued that a road with controlled access acted as a physical barrier: but the lesson had been learned. The Pennsylvania Turnpike is a good example of how the many major toll roads were planned, built and brought into use during this period.

The first meeting of the Pennsylvania Turnpike Commission was held in the office of the Secretary of Highways of the Pennsylvania Department of Highways in June 1937. It was found that, in the passage of the Act by the Legislature, there was no provision for the functioning of the Commission: they were a Commission but they had no funds. They were charged by the Act to finance, construct and operate a super highway running from Middlesex, in Cumberland County, into Irwin in Westmoreland County, across the mountain section of Pennsylvania. However, there was a proviso whereby the Pennsylvania Department of Highways could provide and pay for preliminary design work in preparing for the financing of the turnpike. After many trying months and some disappointments, arrangements were finalised for the turnpike to be financed by an outright grant from the Public Works Administration of 45 per cent of the estimated cost of the turnpike, with the Reconstruction Finance Corporation contracting to take up 55 per cent in the form of bonds bearing 3¾ per cent interest. The formal documents dealing with the financing of the project were signed on 10th October 1938.

The survey of 160 miles had been 75 per cent completed, but there were no construction plans available. With the signing of those documents, the Commission was instructed to invite tenders four days later and to start construction not later than October 27th! On October 14th tenders were invited for earthworks and drainage in Cumberland County, over a length of about ten miles, and the work started on schedule. From that date the design of the turnpike really started. The fundamental objectives were agreed upon as follows:

1. Separation of streams of traffic flowing in opposite directions to reduce the possibility of head-on collisions.
2. The provision of two traffic lanes in each direction to reduce the hazard due to slow-moving vehicles, and with sufficient width to ensure a satisfactory standard of safety for all classes of traffic.
3. The elimination of all highway and railway crossings at the same level so that there would be no cross traffic.
4. The elimination of all fronting commercial developments and local farm entries along the highway with complete control of access.
5. The provision of wide, smooth, stable shoulders, level with the carriageways, to

enable vehicles to park completely clear of moving traffic and to provide elbow room in cases of traffic emergency or mechanical failure.
6. Moderate gradients to give greater safety during icy weather and to enable large, heavily loaded vehicles to travel at higher speeds, so that rear end collisions between fast and slow vehicles might be minimised.
7. The exclusion of pedestrians.
8. The provision of uniform and consistent operating conditions.
9. Specially designed access facilities at reasonably long intervals, with acceleration and deceleration lanes to permit vehicles to enter and leave the main highway with reasonable safety and a minimum of traffic disturbance.

The turnpike was built to a total width of 78 feet, which included dual 24 feet wide carriageways, 10 feet wide shoulders and a 10 feet wide central reserve, except for the tunnels, which only accommodated a single carriageway 23 feet 6 inches wide.

On a visit to the U.S.A. in 1965 I had the opportunity of driving along the whole length of the turnpike and found that work was in progress in duplicating several of the tunnels to take dual two-lane carriageways.

The carriageways of the turnpike were constructed with a 9-inch thick reinforced concrete slab and the total rate of progress achieved by all the Contractors reached a figure of 3½ miles of dual carriageway per day.

The length of 160 miles, costing $71½ million, was completed and opened to traffic on 1st October 1940, only three years after the first meeting of the Turnpike Commission!

Britain's early interest

Although Britain has lagged behind many other countries in the construction of motorways, she has not been without advocates of this type of road since as far back as the turn of the century. In 1900 the then Prime Minister, Mr. Balfour, advocated 'great highways constructed for rapid motor traffic and confined to motor traffic'. His view was endorsed by a large body of public opinion which led in 1905 to the formation of the Road Improvement Association.

A prominent member of the Association's Council was the Hon. John Scott Montagu, M.P., who, in 1913, as Lord Montagu of Beaulieu, was appointed a member of the Road Board on which he served until it was absorbed by the Ministry of Transport in 1919. During this period he was an active campaigner for the construction of motorways, and one of his most interesting plans, produced in 1923, was for overhead roads in congested traffic areas. He continued to hammer this theme with the slogan 'the longer we delay the more we shall have to pay'.[1]

[1] Lady Trowbridge, *John, Lord Montagu of Beaulieu.*

Even in 1923 the desperate state of traffic in the centre of London and other great cities had long been apparent to all students of road transport. With some 4,000 to 5,000 new vehicles being turned out by the factories each week, it was clear that if nothing were done the loss of time and money would become acute.

Lord Montagu turned his attention first to London's problems and as an exercise he attempted an estimate of the cost of doubling the width of three and a half miles of Oxford Street, from the Mansion House to Marble Arch. It was not feasible, he reasoned, to attempt to widen on the south side since it contained too many churches and public buildings. On the north side alone, valuers estimated that the cost of the property on the 19 acres of land involved would be extremely high and the total cost of the scheme would be about £35 million, or £10 million per mile.

Inspired by the success of underground railways, Montagu turned next to the possibility of a tunnel, but his calculations showed that a six-lane tunnel would cost an average of £4 million per mile or £14 million for the total length involved.

Already, Montagu argued, many miles of railway were carried efficiently and economically, if not always aesthetically, above the streets of London, notably the section from Nine Elms to Waterloo, and the lines into Charing Cross and Cannon Street, while Holborn Viaduct showed what an overhead road could do. London streets, he believed, could be similarly dealt with for motor traffic. But reinforced concrete could be used instead of the ugly steelwork of the railway viaducts and would reduce the cost to an estimated £3 million per mile.

He proposed two main overhead routes: one from London Docks to Marble Arch; the other from Surrey Docks to the neighbourhood of Clapham Junction. The roads, for motor vehicle use only, were to be 50 feet wide and carried over long spans at an average height of 100 feet above the streets supported upon great piers, 18 to the mile, of which two per mile would be required for entry and exit along special ramps inside or outside the structure at gradients of 1 in 20 to 1 in 15. He envisaged the remainder being used for offices or flats, thus providing a source of revenue. His final estimate of £36 million for the cost of the roads and the acquisition of the land on which the piers would rest was by no means startling when one considers that the City of New York was then contemplating the spending of £160 million on its roads.

The following year he turned his attention to a plan for a four-lane motorway to the north-west, from London to Birmingham, Manchester and Liverpool. In support of this scheme he rallied the city councils of Birmingham and Wolverhampton and 43 other local authorities concerned, but, as he clearly saw, although revenue from these special roads could be raised from tolls, the capital cost of £15 million could not be financed by bankers' loans as could a going concern. However, when Montagu turned for support to the government of the day and brought forward a Bill to obtain

Parliamentary powers, to construct the road and finance it with tolls, the scheme was instantly set upon and killed by frantic opposition from the railway interests.

Lord Montagu was not alone in the advocacy of motorways. Sir Leslie Scott, in April 1924, introduced a Bill 'to facilitate the construction of motorways and the granting of powers in relation to such ways and traffic thereon'. The move was unsuccessful and 25 years passed before the Special Roads Act, 1949, providing for the construction of motorways, received the Royal Assent.

Although the financial crisis of 1931 and the Second World War caused the postponement of many proposals for major road construction, interest in motorways was still active. In 1936 the Institution of Highway Engineers submitted to the Ministry of Transport a scheme for a national motorway system comprising 51 lengths of road covering 2,826 miles (see Figure 2).

In September 1937 a group known as the German Roads Delegation, numbering 224 members, including Members of Parliament, representatives of highway authorities, highway engineers and others interested in vehicle operation, carried out a tour of inspection of the new German motor road system. The group's report was considered by the County Surveyors' Society and within a month of receiving the report the Society resolved that an adequate study of traffic movement would show that some entirely new roads would be required, but that the number would not be large. These new through roads, with adequate connections to centres of population, would, the Society believed, prove more economical in construction and use than widening existing main arteries to the same standard; segregation of motor vehicle traffic from all other forms would help substantially to reduce accidents; and the construction of motorways for exclusive use by motor vehicle traffic would be considerably cheaper than the construction of roads for all forms of traffic.

The County Surveyors further advocated that the new motorways should be constructed as complete units and not in short lengths; should form a co-ordinated system for through traffic with other arteries of trunk road value; that the whole construction and improvement work should be controlled by a time programme, and existing methods of administration should be improved to speed up works and eliminate irritating delays. In their view new motorways, carefully designed in relation to the landscape, would not injure amenities but, in time, actually improve them. On the other hand, the widening of existing roads, involving demolition of property, would in many cases have the opposite effect.

A Committee of the Society was charged to produce a national plan for motorways which was finalised in May 1938 (see Figure 3). The plan envisaged approximately 1,000 miles at a cost of £50,000 to £60,000 per mile.

The only immediate response to the Society's proposals was a request for information from the Parliamentary Agents to the House of Lords Roads Group. In July 1938, however, the County Councils' Association, on behalf of the Society, decided to forward the proposals to the Ministry of Transport since it was understood that the Minister was investigating the Country's trunk road system.

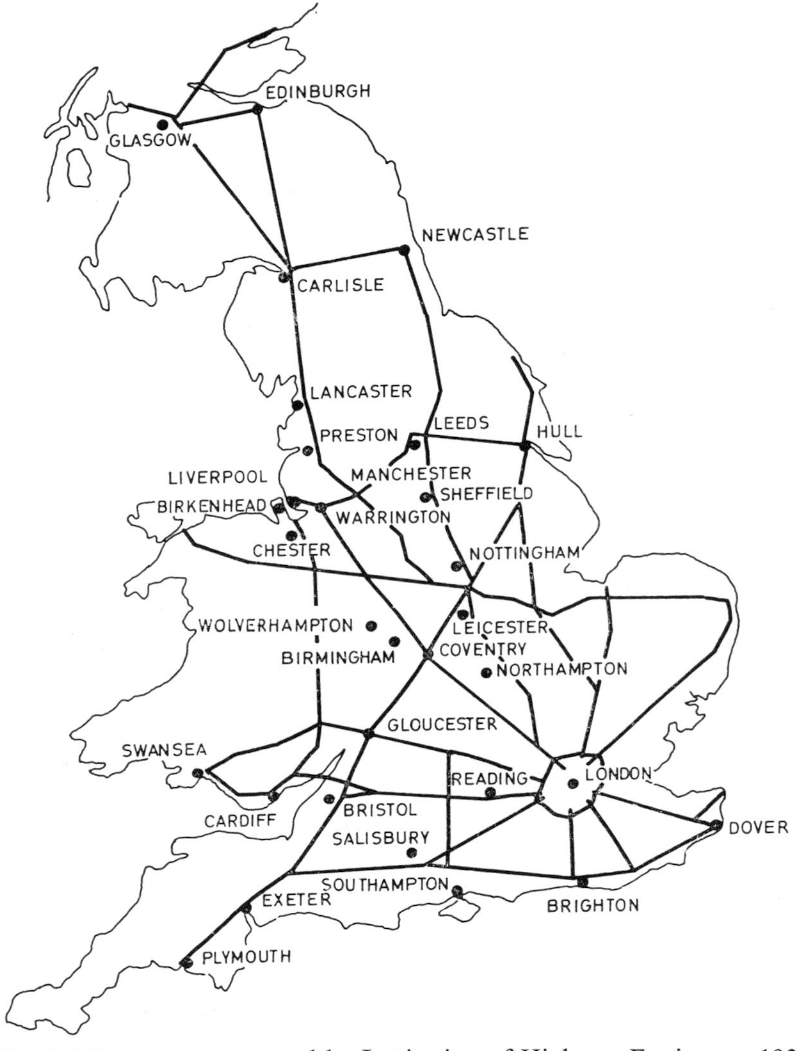

Fig. 2. Motorways proposed by Institution of Highway Engineers, 1936

The Minister of Transport, Mr. Leslie Burgin, following a visit to the German autobahnen, recommended that as an experiment approval should be given to a scheme put forward by Lancashire County Council for the construction of a motorway 62 miles long, running from north to south through the County and parallel to the existing trunk road. Financial restrictions prevented further consideration being given to this project and 27 years were to elapse before this Lancashire section of M.6 was completed.

Fig. 3. National Plan for Motorways, County Surveyors' Society, 1938

In November the Roads Group of the House of Lords and House of Commons received representatives of the County Surveyors' Society and, in May 1939, the Minister of Transport issued a statement on the proposals as follows:

'The German motor roads have excited the admiration–and fired the imagination–of all who have seen them. There can be no doubt that they afford first-rate opportunities for long-distance motor traffic to move fast and with a great measure of safety. The potential advantages are so obvious as hardly to need statement.

'Naturally, then, it behoves us to ask ourselves whether similar methods could not be effectively and economically employed in this Country, both to reduce the appalling toll of accidents and to provide improved traffic facilities for the future.

'It would be hazardous in the extreme to assume, however, that what may be suitable in Germany will necessarily be the best method of achieving that object in this Country. The geographic and economic differences between the two Countries are wide; there is a vast difference in the distribution of population, in the distances and character of country between main centres of population and in the standard and effectiveness of existing communications. For a variety of reasons, geographical as well as social, the cost per mile of constructing a motor road in this Country would be considerably greater than in Germany; more bridges would be necessary, land would be dearer, labour conditions are different. Even the effect on the accident toll is far from being so certain as enthusiasts have assumed; two-thirds of our accidents occur in built-up areas and the traffic in those areas would not be substantially reduced by the withdrawal of the long-distance traffic which would be able to use motor roads. On certain of our main roads, for which figures have been obtained, less than twenty-five per cent of the vehicles travel more than thirty miles.

'The limitation of user of a road to one particular type of traffic would be a novelty in this Country and would require legislation. The question cannot be determined on priority grounds and by reference to vague generalities but calls for detailed examination of the traffic, financial and engineering aspects.

'The subject of the provision of motor roads, i.e. roads limited to use by motor vehicles only, is one which involves many different questions, and in existing circumstances the question of finance is by no means the least important. Whatever may be decided in the future as to the economic value of such roads in this Country, the policy of reconstructing and improving our existing main roads must and will continue, and I can assure you that it is my intention to pursue that policy as quickly and as thoroughly as circumstances and finance will permit.'

In the House of Lords the Select Committee on the Prevention of Road Accidents recommended that an experimental motorway should be built from London to Birmingham but, by the end of 1939, Britain had problems more urgent than roads to think about.

Although the war shelved all road plans for the time being, in 1942 the Institution of Civil Engineers published a report of its Post-war Development Committee[1]

[1] Institution of Civil Engineers, 'Post-War National Development Panel No. 2–Roads', June 1942, Pages 11–13.

which advocated the construction of a 1,000-mile system of motorways on the lines proposed by the County Surveyors' Society in 1938 at an estimated cost of £60 million. This it claimed would obviate the reconstruction of some trunk roads and do away with the need for certain costly by-pass roads.

In 1943 a joint Committee of the County Surveyors' Society and the Institution of Municipal and County Engineers, produced a report which recommended that the motorways should be located in open country, clear of areas of development. Dual carriageways were to be provided, each with two, three or four lanes, 22, 30 or 40 feet wide with marginal strips 12 inches wide on either side, a central reserve 15 feet wide, and outer verges 10 feet wide. A superelevation design speed of 70 miles per hour with curves of a minimum radius of 3,000 feet and a minimum visibility of 600 feet were recommended and for the carriageway construction an 8-inch thick doubly reinforced concrete slab was suggested.

The Institution of Highway Engineers was also looking to the future and issued a memorandum based on their 1936 proposals for the construction of 2,826 miles of motorway. This expressed the opinion that the roads of any Country were the basis of all development and the foundation upon which further progress would depend, and an immediate investigation into the existing highway system as a first step to successful post-war national reconstruction was recommended.

The Ministry of Transport gave renewed consideration to these various proposals. In August 1942 the Chief Engineer, Sir Frederick Cook, prepared a memorandum on post-war planning in relation to motorways which stated the case for motorways and referred to the system proposed by the County Surveyors' Society. It is of interest that a possible route for the London-Birmingham motorway, which was included in this system, had already been investigated by the Ministry.

The memorandum considered whether it was in the national interest for the construction of a system of motorways to form part of the post-war programme but its findings were inconclusive. However, it recommended that where justified, new roads, including by-passes, should be designed for and restricted to the use of motor vehicles.

In May 1943, Mr. Noel Baker announced that the Ministry was preparing plans for the provision of motorways and, in September 1944 a draft design memorandum was produced which detailed the proposed layout including information on types of junction to be used. Access would be completely controlled.

Although, in general, the principles proposed at that time still apply, many changes have been made in dimensions and type of construction. For instance, a major feature was the proposal to provide separate lay-bys for parking rather than the continuous hard shoulder subsequently adopted.

By the autumn of 1944 a national motorway plan was recommended by the Chief Engineer, and after consideration it was decided that preliminary survey work should proceed on a number of priority schemes.

Alas it was to be many years after victory had been won before a start was made on the construction of Britain's first motorway.

Chapter 3

Post-war development

The world awakens

The end of World War II brought new hope to highway engineers throughout the world. Virtually all new highway construction had ceased during the period of hostilities. There was the great opportunity for each country to develop a modern highway system as part of its post-war reconstruction programme.

In those few countries where motorways had been built before the War, fresh ideas based on past experience were being incorporated in future plans. The importance of motorways was also being recognised in countries where little previous interest had been shown. The use of the autobahnen by the Germans had indicated the value of motorways in war. It became clear to many that they could be equally important in restoring a sound peace-time economy.

In the war-scarred countries political and economic demands required priority to be given to such things as housing, industrial development and power production, and the highway programme suffered. It was to be several years before any motorway construction was carried out.

Meanwhile the number of vehicles in all countries increased enormously. Between 1950 and the early 1960's, vehicle ownership on the Continent had more than doubled and it was being confidently predicted that by the 1970's the United States' density, at that time almost one vehicle to three persons, would be reached in parts of Europe.

This was the period when, in all the capitals in Europe, road building pressure groups began arguing for more motorways on grounds of public safety and greater transport economy.

Progress in Britain

In Britain, in September 1944, the Ministry of War Transport produced a draft memorandum on the layout of motor roads, with access restricted to motor vehicles. The preliminary layout for road junctions and the standards given were somewhat similar to those now adopted with two essential differences: the lane and carriageway widths were narrower than the present standard; and there were to be lay-bys at intervals along the motorway instead of the unbroken length of hard shoulder provided today.

In late 1944 the Chief Engineer of the Ministry of War Transport, Mr. A. J. Lyddon, recommended a national motorway plan and it was decided subsequently that initial survey work should be carried out on those motorways considered of first priority.

In June 1945, the suggested design standards for motor roads were put forward by the Chief Engineer of the Ministry of War Transport, Major H. E. Aldington, who had succeeded Mr. Lyddon, in one of three papers read at a joint meeting of the Institutions of Civil Engineers, Mechanical Engineers, Electrical Engineers, Municipal and County Engineers, Automobile Engineers, the Institute of Transport and the County Surveyors' Society.

The first post-war Government statement regarding motorways was made by the Minister of War Transport, Mr. Alfred Barnes, in the House of Commons in May 1946, when giving the pattern of the road programme. This envisaged approximately 800 miles of motorway in the ten-year road construction programme. A map was issued (Figure 4) showing the principal national routes, which included new roads to be constructed as motor roads, together with existing roads which would be generally improved on their present alignment, but with by-passes of the urban areas. The principal routes were located well clear of built-up areas, but subsequent experience showed that the routes must pass closer to built-up areas to enable traffic from the major centres of population to reach the motorway more easily.

On 28th November 1947, a paper entitled 'The Economics of Motorways' was read to the Institution of Highway Engineers by Mr. R. Gresham Cooke, which led to the formation of a Joint Committee of the British Road Federation, the Institution of Highway Engineers and the Society of Motor Manufacturers and Traders to examine the case for motorways and to see whether direct evidence could be produced to show that the construction of motorways would be of financial benefit to the country by the reduction of transport costs. On October 18th, 1948, this Committee submitted a Report to the Minister of Transport and Civil Aviation, which concluded:

'The results of this enquiry show that the construction of motorways is more than justified solely on economic grounds, apart from any considerations of safety, convenience and amenity. The Committee submits that there is no case for questioning the imperative need for action and that motorways should rank with other capital projects which are contemplated in the next few years.

Fig. 4. Ministry of Transport Road Programme, 1946

'The lack of an objective approach to road matters in the past has saddled industry with a heavy burden which today has become a severe handicap to successful competition in the world's markets, on which our maintenance of even a reasonable standard of living depends. We cannot afford inefficient road communications and, though the construction of an adequate system of motorways will not be a panacea, it will provide the greatest return to be obtained from the expenditure of national effort and resources. In this lies the justification for motorways being accorded the highest priority of all major schemes of road development.'

On 30th January 1948, another paper by Major Aldington, entitled 'The Design and Layout of Motorways' stressed that motorway planning must be given its proper place in the road network and that the ideas should be sufficiently flexible so as to be capable of modification in the light of new developments.

Not until 1949 were the necessary legal powers obtained in the Special Roads Act 1949 to restrict motorways to the exclusive use of certain motor traffic. This Act legalised both the construction of special roads and the conversion of existing roads into special roads, to which access would be restricted, and which would be used only by certain specified classes of traffic. The Act also gave the necessary powers to alter or stop up any side roads, public footpaths, or private accesses which crossed the route of the special road. It also prohibited statutory undertakers, with the exception of the G.P.O., from laying services within the area of the special road.

Worcestershire was the first County to have the line of a motorway established by means of an Order when in 1949, 'The North of Twyning–North of Lydiate Ash Trunk Road Order' fixed the route of the 28-mile length between Brockeridge Common and Lydiate Ash, now part of the Bristol-Birmingham Motorway, M.5.

In common with other motorway routes, preliminary location work on the Ross Spur Motorway extended over several years but, in 1953, the proposal received a new impetus. A Committee, headed by Lord Lloyd, and formed to study the communications of South Wales, reported and strongly recommended the construction of a route linking the industrial areas of the Midlands and South Wales. This route was suggested between the Worcester-Tewkesbury Road, A.38, north of Tewkesbury, passing through Worcestershire, the north-western tip of Gloucestershire, to Ross-on-Wye in Herefordshire, south of the Malverns.

In February 1953 the Minister of Transport and Civil Aviation, Mr. A. T. Lennox-Boyd, announced the inclusion in his road programme, for authorisation in 1956–9, of a scheme for the construction of the Stretford-Eccles Motorway in Lancashire, about six miles long with a high-level bridge at Barton, over the Manchester Ship Canal. It was initiated by the Lancashire County Council, which recognised the urgent necessity of proceeding with the construction of a motorway to replace the existing inadequate Class I road and, at the same time, to eliminate the repeated frustrating and lengthy delays to traffic which occurred at the swing bridge over the Manchester Ship Canal.

In May 1953, Mr. Lennox-Boyd intimated his intention of making a Scheme under the Special Roads Act, 1949, to fix the centre line of the eight-mile-long Preston

By-pass as part of the M.6 motorway. Then, on 8th December 1953, the Minister included Preston By-pass and Lancaster By-pass as motorways in the Government's road programme, and on the same date the Home Secretary announced that the Ross Spur Motorway was also included.

On 2nd February 1955, Mr. J. A. Boyd-Carpenter, the new Minister of Transport, included the London-Yorkshire Motorway and the Birmingham-Preston Motorway in the expanded programme of schemes amounting to £147 million over the next four years. On 12th June 1956, construction work was started on the Preston By-pass section of M.6, on 10th April 1957, the Stretford-Eccles By-pass section of M.62, and on July 5th 1957, the Lancaster By-pass section of M.6.

On 24th March 1958, work started on the construction of the London-Birmingham Motorway, M.1, and on 31st May 1958, the Ross Spur Motorway, M.50. Although the layout of the Preston and Lancaster By-passes provided ultimately for three-lane carriageways, M.1 was the first motorway actually to be built with three-lane carriageways. The layout of the Ross Spur Motorway, like the Stretford-Eccles By-pass, M.62, only provided for dual two-lane carriageways.

These were the first five motorways to be started in Britain.

In the meantime, on 22nd July 1957, the Minister, Mr. H. A. Watkinson, announced a larger programme for the next four years, totalling £240 million. The five main projects were: (i) improvement of the Great North Road from London to Newcastle; (ii) a motorway from London to the North-West; (iii) better communications between the Midlands and South Wales; (iv) better roads to the Channel Ports; and (v) a new outlet from London to the West. In this programme, the London-Birmingham Motorway, M.1, served as the southern part of both the London to the North-West and London to Yorkshire Motorways. The other motorways in the programme, which were included in the five main projects, were the Birmingham-Bristol Motorway, M.5, southwards to its junction with the Ross Spur Motorway, the Medway Towns Motorway, the South Wales Radial Road between Chiswick and Slough By-pass, the Slough and Maidenhead By-passes and the Severn Bridge.

On 5th December 1958, the Prime Minister, Mr. Harold Macmillan, officially opened the Preston By-pass. This was followed, on 2nd November 1959, by the opening by the Minister of Transport, Mr. Ernest Marples, of the 72 miles of the London-Yorkshire Motorway, M.1, between Aldenham and Crick. On 11th April 1960, the Lancaster By-pass was officially opened by the Chancellor of the Duchy of Lancaster, Dr. Charles Hill.

In 1959 the Provisions of the Special Roads Act, 1949, were incorporated in the Highways Act, 1959, the relevant sections of which are referred to later in Chapter 4.

On 3rd May 1960, Mr J. F. A. Baker, then Chief Engineer of the Ministry of

Transport, presented a paper 'The General Motorway Plan' to the Institution of Civil Engineers.[1] Another interesting paper was given by Mr. H. N. Ginns, Deputy Chief Engineer of the Ministry of Transport on May 18th, 1966.[2]

Table 4 shows the mileages of motorways completed each year up to 1968 in

Table 4
Mileage of motorways completed each year in England and Wales

	Motorways completed in miles	
Year	In year	Cumulative total
1958	8	8
1959	73	81
1960	45	126
1961	21	147
1962	50	197
1963	95	292
1964	8	300
1965	75	375
1966	53	428
1967	97	525

England and Wales, whilst the position at the end of 1968 is shown in Figure 5. By this time about 550 miles of motorway were in use, a further 160 miles were under construction, and the line of the route had been fixed for an additional 300 miles of motorway.

Many organisations and individuals have tried to further the construction of motorways in Britain and none more than the British Road Federation and the Roads Campaign Council, who have made unceasing efforts to draw attention to the essential role of motorways in the economic and social welfare of the nation.

The aim of the Government is to build 1,000 miles of motorways by the early

[1] J. F. A. Baker, 'The General Motorway Plan'. *Proc. I.C.E.*

[2] H. N. Ginns, 'English Motorways–Development and Progress'. Public Works and Municipal Services Congress 1966.

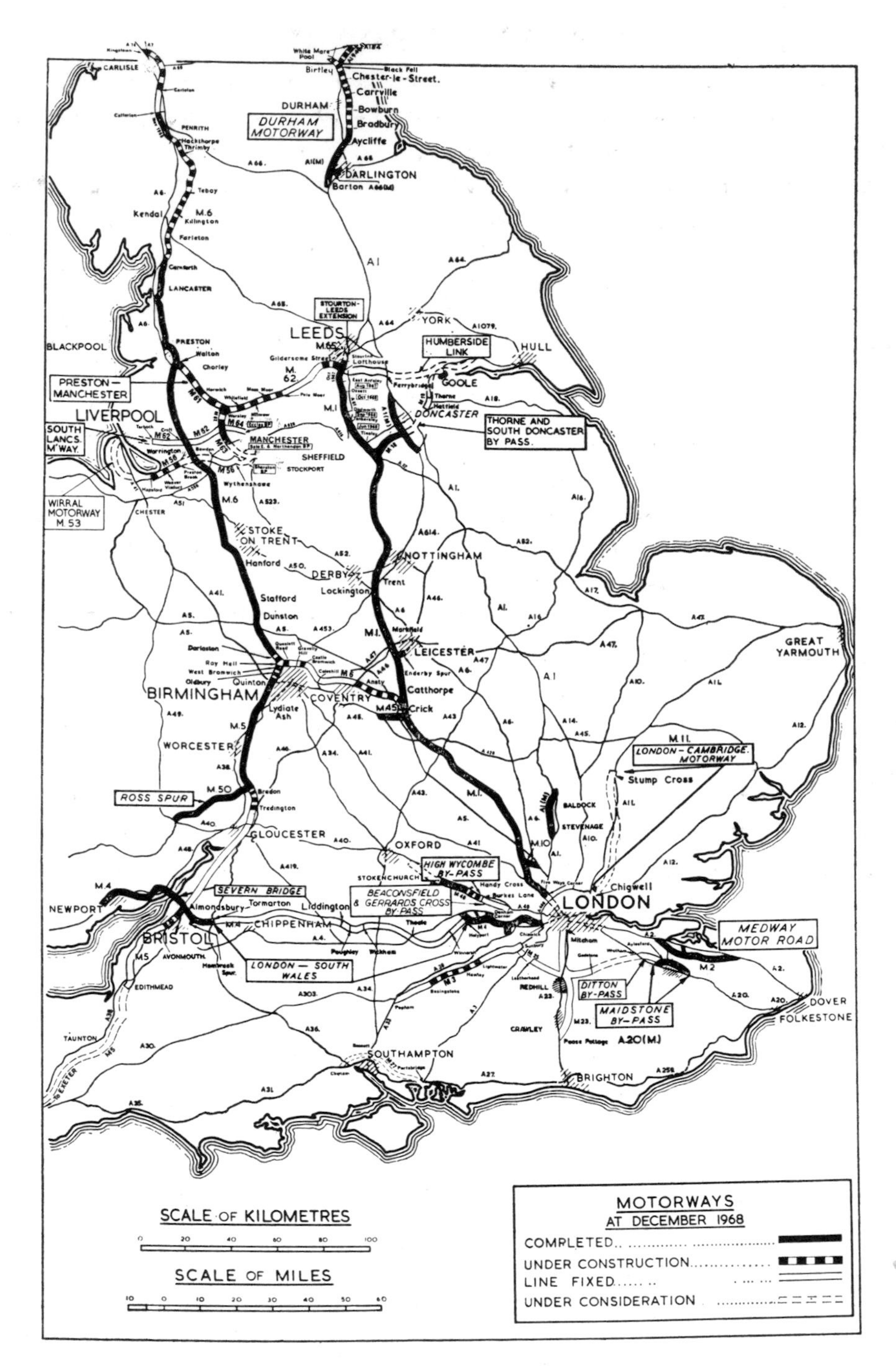

Fig. 5. Motorways in Britain, 1968

1970's and, although Table 4 may raise doubts as to the feasibility of achieving this aim, I am of the opinion that with the increased attention now being paid to motorway construction, this target will, in fact, be achieved. This was one of the objects of the formation of the Road Construction Units in 1967 – a proposal which was initiated by the then Minister of Transport, Mrs. Barbara Castle, and which has been implemented by Sir William Harris, the Director General of Highways.

Germany

In West Germany there is no doubt that one of the reasons for the rapid economic recovery in the immediate post-war years was the existence of the autobahnen. Of the 2,500 miles opened to traffic before the War in Germany as a whole, over half is now in West Germany. Valuable experience, therefore, was available to the West German Authorities in planning their future programme of construction.

The design speed was reduced from 100 miles per hour to 75 miles per hour. The long straights which were a feature of the pre-war autobahn were to be eliminated from the alignment of the future motorways and replaced by a series of curves. To cater more for local traffic the spacing of the interchanges was to be reduced and at the interchanges longer acceleration and deceleration lanes were to be provided. The length and steepness of gradients were to be limited to 1½ miles and 1 in 22 respectively. The width of the hard shoulders was to be increased to 8 feet and guard rails were considered to be indispensable at the top of embankments. The importance of adequate servicing facilities was recognised and the spacing of service areas reduced from 40 miles to 20 miles. Telephones were to be installed at two-mile intervals.

These revisions in standards had been developed from a detailed examination of the operational requirements of future traffic and from the accident records of the earlier motor roads. It was found that the accident frequency was only 43 per cent of that on existing all-purpose main roads and the fatality rate only one-third as high. Even so, since these figures were considered to be excessive, it was realised that several design features required investigation.

Although heavy goods vehicles accounted for only 26·5 per cent of the vehicle mileage run on the motorway system, they were involved in 41 per cent of the accidents. It appeared, therefore, that the relative speed differential between classes of vehicles was a cause of accidents. A reduction in the steepness of gradient was thus called for, and where this could not be reasonably achieved, additional traffic lanes should be provided for the slow-moving climbing vehicles.

Over 25 per cent of the accidents occurring on the motorways arose from vehicles breaking down where there was an inadequate width of shoulder for them to draw off the carriageway. Of these, 70 per cent occurred in conditions of fog, heavy rain, snow or darkness. Many accidents were attributed to drivers falling asleep on long straight stretches, particularly in the case of slow-moving vehicles. Investigations showed that a road composed of continuous curves, which offers the driver a constantly changing view, gave rise to increased traffic safety.

Careful attention was accordingly given to all these factors in the design of post-war motorways.

By 1965 an expenditure of over £300 million had increased the total autobahn mileage to 2,000. The third four-year Plan for 1967–70 provides for the construction of a further 700 miles, requiring an acceleration of the annual building rate from 75 to nearly 200 miles a year.

The importance of road construction in the context of long-term expansion has long been recognised in West Germany. Between 1960 and 1964 the number of cars increased by 82 per cent but the road network by only 12 per cent. The number of vehicles is estimated to increase by 110 per cent in the ten-year period 1965–75. Future plans provide for 4,425 miles of autobahn to include many urban motorways and connections from the North Sea ports to the main systems.

Italy

In few countries is a system of motorways likely to bring about so great a social and economic revolution as in Italy. Its shape and the contrast between the industrialised north and the backward south has been her principal problem for many years. The original 'autostrade', built as high speed toll roads in the 1920's and 30's, were not of any great direct benefit to the nation's economy. The new network, however, is already having a remarkable effect throughout the country.

The current motorway programme was initiated in 1955 and by 1966 over 1,100 miles were in use and a further 1,000 miles either under construction or in the advanced planning stage. With future schemes already programmed, it is expected that a network of 3,000 miles will be in operation by 1970.

Ultimately there will be two principal north-south autostrade, the completed Autostrada del Sole or Highway of the Sun, from Milan to Naples, which will eventually run further to the south, and the Bologna to Bari autostrada along the Adriatic Coast. The two will be connected in the north at Bologna and, in the south, by a Naples-Bari Autostrada.

The completed 470-mile length, opened in October 1964, reduced the distance between Milan and Naples by 60 miles and cut as much as four hours off journey times. Apart from its value to industry, the spectacular nature of the road and the country which it traverses make it a popular tourist attraction.

The State Holding Company, IRI (Institute for Industrial Reconstruction), is responsible for nearly half the Italian programme. The Company constructs and operates a group of selected toll motorways in such a way that the more profitable ones compensate for those which are less profitable but which seem equally useful in terms of social and economic benefit.

Another method is for the State to award concessions and make grants to private companies for certain toll roads. The concessionaire then issues bonds to cover the remaining costs, repayable over a period of 30 years, after which time the State takes over the roads.

A third alternative is by the award of concessions without a state contribution. With this method the period may be very much longer and extend to 70 or 80 years, as in the case of the Turin-Milan Autostrada and the Great St. Bernard Tunnel.

In addition a number of motorways, free of tolls, are being constructed in the south which are the direct responsibility of A.N.A.S., the National Roads Board, the central body which has the overall control of all Italian roads.

France

In relation to its size and vehicle density France showed remarkably little interest in motorways in the immediate post-war years. This was more a question of policy to assist the railways rather than of simple indifference.

In 1953 motorways in use totalled 33 miles and, by 1963, 216 miles, of which 66 miles were opened during that year. Initially all motorway construction was centred around Paris with the emphasis on the Autoroute du Sud, the route linking the capital with Lyons and Marseilles. Latterly, work has been carried out on the A-1 motorway between Paris and the important industrial city of Lille to the north. This involved the construction of a section in cutting between retaining walls through the densely populated northern Paris suburbs. On this heavily trafficked length which connects with the international Le Bourget Airport, a minimum number of eight traffic lanes has been provided with ten on certain sections.

The opening of the last section in 1965 gave France its longest uninterrupted motorway length of 58 miles. Construction was financed by a company owned jointly by the State and private interests. Tolls are charged, except on the first 12-mile length out of Paris which, because of its importance as an access road to the City, is free. The A-1 is regarded in France as the Common Market artery since it will connect Paris with Belgium, thence Germany and also the Dutch ports.

In the early 1960's there was an awakening to the need for an improvement in the rate of construction. A five-year plan, published in 1965, provided for the construction of 620 miles of motorway as an addition to the 400 miles then in use. The plan included the completion of the north-south motorway connecting Lille-Paris-Lyons-Marseilles-Nice, and a motorway through Lorraine to connect with the German autobahnen.

On a long-term basis France is planning to provide some 4,500 miles of motorway by 1985 at a rate of construction of 190 miles per year. By then, the vehicle population is expected to treble from its 1965 figure of eight million, which represents a vehicle density of one car for every 2·8 persons.

Motorway building in France is financed from three different sources: a special road fund which receives 12 per cent of a special fuel tax and provides almost half the money required; loans repaid out of tolls levied at approximately 2*d.* per mile outside the cities which provide a third; and Government contributions which provide the remainder.

Belgium

In Belgium the Brussels-Ostend Motorway was actually started before World War II, but the whole 75 miles were not completed until 1956. It forms part of a network of 625 miles of autoroutes devised in 1949, when a 15-year plan was prepared.

By 1965, 180 miles of motorway had been opened to traffic and the plan had been extended to include 850 miles by 1980. One of the most important roads in the network is the 105-mile section of the E.3 which will form part of the international road system linking Lisbon to Stockholm. Whereas in Belgium motorways are normally financed from central Government funds, special arrangements were made for this particular road. A company, formed by the local authorities through whose area the road is routed, was granted a Government concession to construct, maintain and operate the road for a period of 30 years. The construction is financed by loans from private banking and finance houses. Usually the concessionaire of a motorway is indemnified by a toll paid by the road users but in this case, a State contribution calculated on the basis of a rate for each vehicle using the motorway has been substituted. The user pays no toll and the Government's financial commitment is postponed until the road is open to traffic. The debt is then spread over 30 years.

Holland

In Holland it had been realised in the 1920's that the improvement of the road network was a vital necessity, especially in the western part of the Country, where the land has been largely reclaimed from the sea. Here there were difficult problems not usually met with elsewhere, the area consisting of an irregular patchwork of 'polders', varying in size up to about 75 square miles and completely surrounded by dykes which carry the roads.

The old road network served its purpose well up to the coming of motor traffic, but it soon became evident that, because of their sinuous alignment, inadequate width, and unstable sub-soil condition, it would be impracticable to adapt the dykes for modern traffic. It was decided, therefore, that a new system of primary roads was necessary from which agricultural and certain other classes of traffic would be excluded. The number of interchanges with other roads was to be kept to a minimum.

It was only a step further to give the most important of these roads the character of motorways by controlling access from adjoining land, preventing development on adjoining land, eliminating intersections at one level and separating opposing traffic with a central planted strip. Where the volume of traffic did not yet justify immediate construction as motorways, the work was to be carried out in stages. Initially only one carriageway would be constructed and intersections at the same level would be permitted. Subsequently improvement to a full motorway was undertaken.

In the eastern part of the Country the old roads were much straighter and easy to widen. In addition the ground conditions were better, the sub-soil being largely sand. It was at first thought that the existing roads could be improved with by-passes provided around towns and villages but the mingling of fast, long-distance traffic with agricultural and local traffic, and the many accesses and intersections produced unsatisfactory results.

By the start of the War some 70 miles of motorway had been built, and, although not large compared with the mileage constructed in the U.S.A. or Germany, Holland's population is only twice that of Lancashire and there are only 75 per cent of the number of motor vehicles. By 1966 the mileage had increased to nearly 500 miles, either completed or under construction.

Sweden, Denmark and Norway

The first motorways in Sweden, Denmark and Norway were opened in 1953, 1957 and 1965 respectively. Further advances have been made since then and, by 1966, Sweden had 178 miles either open to traffic or under construction and Denmark 152 miles. Whereas in Sweden the motorways are concentrated around the larger cities, in Denmark they form important connections from Copenhagen to the North Sea ports and the ferry crossings to Germany. I have been impressed with the magnitude of the bridges in Denmark, both under construction and in the programme, particularly for a country of its size and with a population of less than five million.

Austria

The geographical location of Austria demands the provision of a first-class road system for its future progress. As a bridge between north and south, and a link between east and west, by establishing favourable conditions for international through traffic, it is felt that the more it succeeds in leading the life lines of Europe through its markets, the greater will be its economic growth.

The Salzburg-Vienna Motorway, 190 miles long, was virtually complete in 1965. It is at the eastern end of the longest single length of motorway in Europe from Antwerp in Belgium to Vienna, a distance of 770 miles.

With a total of 230 miles of motorway either open to traffic or under construction in Austria, plans are well advanced for a motorway route from Vienna south and west to connect with Italy and Yugoslavia. A further route through Innsbruck connecting the Brenner route with Italy to Munich in Germany is planned to provide the bridge between north and south Europe.

Switzerland

Switzerland constructed its first length of motorway in 1960 and plans to complete a network of 450 miles by 1985. A total length of 190 miles had been opened to traffic by 1965.

Other European countries have also constructed motorways and have ambitious programmes for the future. The progress made between 1955 and 1965 was such that the mileage in use had more than doubled. If Governments keep to their forecasts the 1965 mileage, as shown in Figure 6, will have multiplied by three again by the early 1970's and an 11,500-mile network will extend from the Pyrenees to Vienna and from the toe of Italy to Denmark.

Apart from Europe two countries in different parts of the world are moving ahead rapidly with programmes of motorway construction.

South Africa

In the Capetown area of South Africa, the basis of the system consists of radial freeways. The first three miles of the freeway from the north were constructed in 1959. With dual three-lane carriageways provision was allowed for future widening to dual four-lanes. The extension northwards to a total length of 38 miles and with 16 interchanges is to be completed by 1970. Similar developments have taken place in the construction of parkways running east and west from the City. Experience has shown that the average traffic speed has increased to 45 miles per hour compared

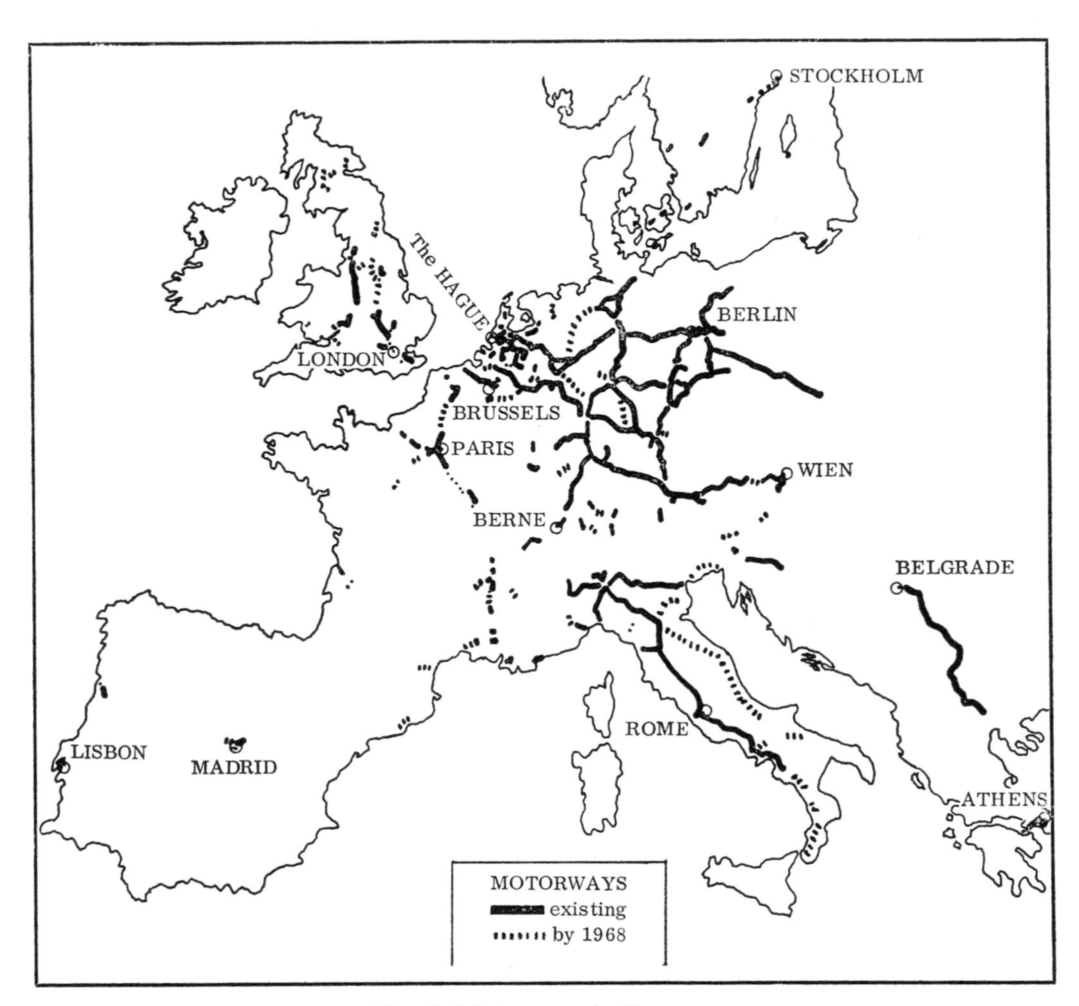

Fig. 6. Motorways in Europe

with 20–25 miles per hour or less on the old routes. Accidents have been reduced by between one third and one fifth.

Motorway systems are also being planned in Johannesburg, Pretoria, Durban and Port Elizabeth. The National Transport Commission, which functions as a National Road Board, is going ahead with proposals for a giant freeway system as a replacement for all national roads in the Country.

Japan

In the ten years from 1955 to 1965 Japan has emerged as one of the leading industrial nations, with its economic growth rate outpacing that of the largest western nations, including the U.S.A. and Great Britain. A latecomer in the field of long distance motorway construction, it has only recently embarked on a series of projects to handle the rapidly increasing volume of private and commercial vehicular traffic.

Its first effort, the 118-mile Meishin Expressway between Kobe and Nagoya, was completed in June 1965 and cost approximately $331 million. The extension of this road to Tokyo, a length of 215 miles, began in April 1965 and scheduled for completion in March 1969, will cost $950 million. It is estimated that the system, passing through the most heavily industralised areas in Japan, will halve travelling time between Tokyo and Kobe by 6½ hours.

Other major motorways were projected in a five-year Road Programme covering the years 1964–8 at an estimated cost of $11,400 million.

Meanwhile in North America great strides have been made in major road construction.

Canada

The first motorway to be built in Canada was the 'Queen Elizabeth Way' between Toronto and Hamilton in the Province of Ontario. Completed in 1939, it has a length of approximately 60 miles of dual two-lane concrete carriageways with a central reserve 60 feet wide.

Between 1945 and 1955 further motorways were constructed radiating from Toronto, and increasing the total mileage to about 200. This included the first sections of Highway 401, 'The Kings Highway', which now extends for 550 miles across Ontario from Quebec to the U.S. border at Windsor.

West of Toronto in the more rural areas, 'Highway 401' was not constructed to full motorway standard. Built 15 to 20 years ahead of its time, in order to encourage industrial development, traffic was very light when this section was opened. The greater portion had dual carriageway construction and major roads were carried over by means of bridges, but direct accesses from some minor roads were allowed, to reduce the initial cost. A programme is to start shortly to build overbridges at these access points and convert the road into a completely controlled-access motorway.

During the ten years to 1965, motorway mileage in Canada increased to over 1,200 miles mainly concentrated around the principal cities.

Seven of the ten Provinces now have motorways, with Ontario the undisputed leader in the field with nearly 400 miles of motorways with complete control of access.

In British Columbia, 113 miles of motorway were built radiating south and west from the City of Vancouver. In Alberta, 49 miles were completed with an additional 170 miles under construction, all of which are located between Calgary and Edmonton. Approximately 40 miles of motorway were constructed leading north and west from the City of Regina in the Province of Saskatchewan. Motorways were also built in the city of Saskatoon.

The Province of Quebec is also making rapid progress with the section of the Trans-Canada Highway planned as a motorway between Quebec City and the Ontario border, while other motorways are under construction radiating from Montreal. The Province expects to have 1,000 miles of motorway by 1970.

U.S.A.–The National System of Interstate and Defense Highways

The origin of the Interstate Highway System can be related specifically to a formal report made to Congress in 1939. It had become evident that the dual-purpose conventional primary highway could no longer safely and efficiently provide for large volumes of express travel between major traffic or distribution centres. Roads specifically designed and dedicated to moving high density, high speed traffic, free from the duty of serving adjacent property, were indicated.

Following a further report to Congress in 1944, the Federal Aid Highway Act was enacted, designating a third Federal Aid system to supplement the existing primary and secondary systems. The National System of Interstate and Defense Highways was to be planned to connect by routes as direct as was practicable, the principal metropolitan areas, cities and industrial centres, to serve national defence and to connect at suitable border points with routes of continental importance in the Dominion of Canada and the Republic of Mexico.

Not until 1956, however, were any specific appropriations made by Congress. The Federal Aid Highway Act of that year was a unique piece of legislation by virtue of its magnitude and its widespread effect. By undertaking to provide Federal Aid on a basis of 90 per cent of the cost, it made possible the implementation of a National system of 41,000 miles of Interstate highways (Figure 7), principally motorways.

The U.S. Bureau of Public Roads, on the basis of forecasts made by the State Highway Departments, estimated that between 1956 and 1976 the total number of vehicles registered would increase from 65 million to over 113 million, while the mileage travelled would rise from 623,000 million to 1,200,000 million vehicle-miles

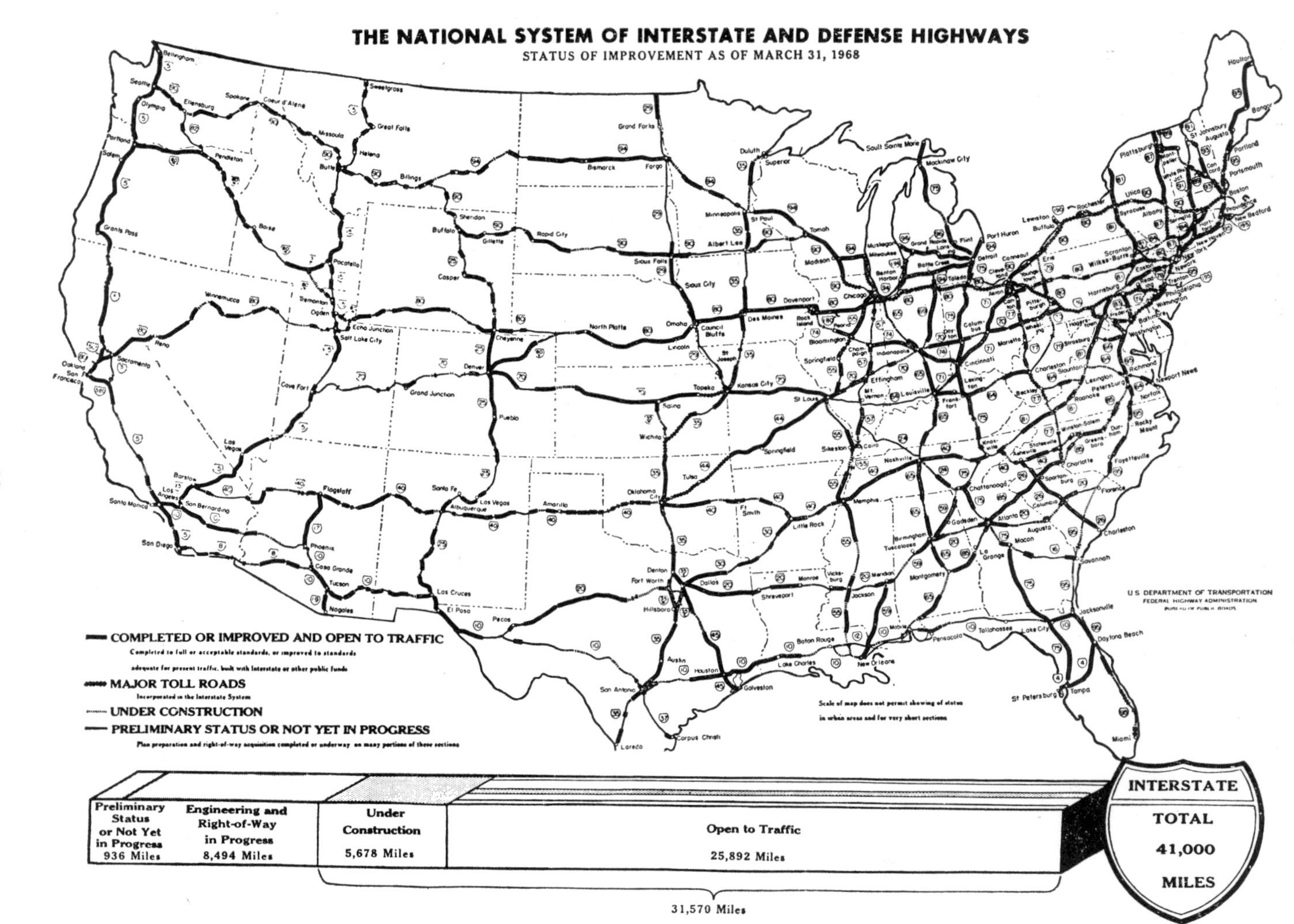

Fig. 7. The National System of Interstate and Defense Highways–U.S.A.

annually. It is predicted that the System, while representing only one per cent of the total U.S. road mileage, will carry 285,000 million vehicle-miles in 1976 or nearly 25 per cent of the total.

The estimated total cost, approximately $50,000 million, makes it the largest project of its kind ever undertaken; $45,000 million will be met by the Federal Government and the remaining $5,000 million provided by the States. The Federal money comes wholly from the Highway Trust Fund, which derives its income from the following highway user taxes:

Motor fuel	4 cents per gallon
Lubricating oils	6 cents per gallon
New trucks, buses and trailers	10 per cent of the manufacturer's sale price
Parts and accessories	8 per cent of the manufacturer's sale price
Vehicle tyres and tubes	10 cents per lb. weight
Heavy vehicle use	One dollar per 1,000 lb. annually on the gross weight of vehicles of more than 26,000 lb. gross weight.

Prior to 1956, Federal Aid Highway funds were appropriated from the general treasury with no direct relationship between Federal Aid for highways and Federal taxes on highway users. The Highway Trust Fund, however, is inviolate and any under-expenditure in a particular year is carried forward for future use.

Continuing in the tradition of the older Federal Aid programme, the building of the Interstate system is accomplished by means of the well-established State-Federal Partnership. The routes comprising the System are selected by the State Highway Departments with the approval of the Bureau of Public Roads, which has the advice of the Department of Defense on certain features of the programme.

The small town is usually skirted by the Interstate route to which it is connected by access roads and interchanges. The towns do not suffer any disruption and as there are no service facilities on the Interstate system, drivers may leave the main route to take advantage of the facilities available in them. In contrast, the large urban areas will be served both internally and externally by a system of freeways of which the Interstate routes may form the major framework.

The initiative in carrying out the programme is basically with the State Highway Departments, who select and design projects to be undertaken each year, acquire the necessary land or 'right of way' and supervise construction. The title of the Interstate routes, as for all Federal Aid projects, rests in the States who accept full responsibility for their maintenance.

The strength of the partnership of the States with the Federal Authorities is largely attributable to the activities of the American Association of State Highway

Officials,[1] which in co-operation with the Bureau of Public Roads, a member of the Association, has developed the various technical standards, policies and specifications for the Interstate Programme.

The principal design features of the Interstate System are motorways of four to eight traffic lanes, dual carriageways, hard shoulders, full control of access, flyover crossings and acceleration and deceleration lanes at interchanges.

When completed in 1975 the System will connect and serve some 90 per cent of cities with populations of over 50,000 and many smaller towns and cities as well.

It is expected that the System will save some 8,000 lives a year after completion, taking into account lives saved on both the Interstate System because of design improvement and those saved on older highways relieved of some traffic. A comparison made of the fatality rates on parts of the Interstate System open to urban traffic, with older roads in the same traffic corridors which formerly carried most of the present Interstate traffic, showed the combined rate for the Interstate and other roads dropped to 3·9 per 100 million vehicle-miles from 5·1 on the older roads, while on the Interstate routes the fatality rate was only 2·6.

A study carried out by the Bureau of Public Roads has shown that control of access is the single most important safety factor. The other contributing factors included wide central reserves, dual carriageways, easy curves and gradients, and long sight distances.

It has also been shown that time and not distance is the most important travel consideration. Drivers will in general choose the route offering the shortest journey time even if it means travelling a greater distance, with savings to the user in reduced operating costs, lowered fuel consumption, wear and tear, and depreciation. Direct benefits to users of the Interstate System are estimated to reach $11,000 million per year in 1975.

There are, of course, many indirect benefits accruing from the System. Considerable industrial and residential development has taken place around many of the interchanges. Land values have risen and there has been a rapid increase in the prosperity of areas served directly by the System. In the rural areas, improved accessibility has made it possible for land to be converted to more productive use.

Of the 41,000 miles of Interstate routes 13 per cent are in urban areas and 87 per cent in rural areas. Some 80 per cent will be built in completely new locations and six per cent of the total mileage is in existing toll roads, bridges and tunnels. A total number of 12,100 interchanges involving 18,600 individual structures will be required. A further 35,000 structures will provide crossings for roads, railways, rivers and

[1] American Association of State Highway Officials, *The First Fifty Years 1914–1964*.

streams. The average spacing of interchanges in rural areas is $4\frac{1}{2}$ miles and less in urban areas.

By the end of March 1968, 12 years after the System got under way, 23,755 miles had been opened to traffic in 49 of the States and in the District of Columbia. The State of Alaska is not involved in the programme.

The ultimate effects of the Interstate System cannot be fully appreciated at this stage. The greatest peacetime public works programme ever undertaken may well be one of America's most beneficial public investments in both social benefit and economic return.

U.S.A.–Toll Roads

The programme did not preclude the separate States from proceeding with their own programmes and financing them by tolls or otherwise.

Heavy traffic demands and inadequate funds spurred toll road construction in the 1950's, mainly in the eastern and mid-western States. By means of legislation, an Authority was established to finance and construct the toll road as an income-producing project. The Authority is a public corporate body with adequate powers and capable of attracting capital from banking sources.

Financing by the issue of revenue bonds requires co-operative teamwork between engineers, bankers and lawyers to ensure that the many factors involved are accurately assessed. These include careful estimates of the cost of the project, the gross revenue, and the cost of operation and maintenance. All this information has then to be applied to the terms of the contract with the bond-holders, with safeguards as to the time of the completion of construction, the fixing and collection of tolls and assurances as to the proper and economical operation and maintenance.

In some cases, however, the States have found it necessary or desirable to pledge other revenues as a supplement to tolls: in effect, to provide a subsidy in order to assure the solvency of the project.

The success of the earlier toll roads, such as the Pennsylvania Turnpike, had however justified those who had pleaded their necessity and economic value. When the first 160-mile length was proposed it was estimated that the cost would be recovered from tolls over a 30-year period. In fact it could have been freed in a much shorter period but instead, the system was extended and now totals 470 miles. During the 10-year period after the first section was opened, 2,500 million miles had been covered by 25 million vehicles yielding an income to the Turnpike Commission of $42 million in tolls.

It is clear that where express highways cannot be provided as free roads, both private motorists and commercial vehicle operators are willing to pay tolls in order to use them. There are, however, many engineers and public officials in the U.S.A. who are opposed to the principle of toll roads. It is argued that only a few routes are suitable for toll financing, such as the heavily trafficked connections between large cities and, therefore, the method does not offer a national solution to the urgent need for freeways. For toll collections to be economic, access and egress points must be limited to a few points where traffic volumes are large. This may prevent the road being used for short journeys, particularly in the urban areas.

In addition to the payment of interest on the capital, the cost of toll collection can be quite high and both have to be borne by the user. It is also argued that the cost of construction is higher than that of a freeway because of the more complex nature of the interchanges necessary to lead traffic through the toll collecting stations. There is also evidence that, in some instances, the Toll Road Authorities have paid excessively high rates for the acquisition of land in order to complete the construction early and obtain revenue.

The main objection, of course, is that users are paying tolls in addition to petrol and other forms of vehicle taxation. They are paying for free roads which they do not use.

The body of opinion which favours toll roads points out that, where feasible they can be provided quickly, and also because the projects must be self-liquidating only those roads will be constructed which are economically justified, and further that the Authority concerned will design and construct in the most economic way possible.

It is also pointed out that, where a toll road carries mainly long distance through traffic, there is no burden on the State through which it passes. For example, it would be argued that the State of Connecticut, which lies across the main route between New York and Boston, should not bear the cost of providing a road for the benefit of traffic travelling mainly between two major cities, both outside the State, and therefore, a toll road would be more equitable. The high level of Federal Aid, however, for the Interstate routes defeats this argument.

The New Jersey Turnpike

The story of the New Jersey Turnpike[1] illustrates a remarkable achievement and provides a good example of a successful toll road. Geographically New Jersey is one of the smaller States but ranks high in terms of population and importance. Because of its location it has a concentration of traffic greater than anywhere else in the U.S.A. or, in fact, anywhere else in the world. It is a corridor lying between the

[1] Paul L. Troast, and Others, 'New Jersey Turnpike', New York, American Society of Civil Engineers, January 1952.

populous areas of New England and New York in the north-east, and the States lying south and west. It also connects New York and Philadelphia, the first and fourth largest cities in the country, and must therefore accommodate a considerable volume of through movement in addition to the State's own traffic.

In 1946 New Jersey's Governor realised that extraordinary measures would be necessary to deal with the predicted post-war traffic increase. Since at that time the State Highway Department had only some $30 million available annually for new highway construction, he personally sponsored legislation to authorise the setting-up of the New Jersey Turnpike Authority and, in 1948, the enabling Act was passed.

The Authority was established in March 1949 and consists of only three members, all businessmen without any political or engineering experience invited personally by the Governor.

The immediate need was seen as a new major road to serve as a 'backbone' for the State's entire highway system, extending southwards from the George Washington Bridge to connect with the Delaware Memorial Bridge (then under construction across the Delaware River). It would also serve all the main centres of population in the State along its length of 118 miles.

The economic feasibility of the project was evaluated in three stages. First, an estimate was made of potential traffic based on existing conditions, which were assessed by means of a number of origin and destination surveys carried out at twelve different stations covering the competitive routes. A representative sample of traffic flow was established with the help of the State and local police and a number of interviewers. In addition to obtaining this information, total traffic counts were undertaken which were classified into traffic types, and the State registration plates of the vehicles were also taken into account. These data were analysed to determine the number of vehicles which would use each part of the turnpike.

Second, the advantages offered by the proposed turnpike, as compared with existing highways, were evaluated to estimate traffic that might be diverted to it. Specific advantages were clearly the saving in time, or distance, or both, and easier and safer driving due to the absence of congestion and the elimination of delays at intersections.

Third, an estimate was made of the optimum toll rates to be charged on each section of the turnpike. Experience of earlier turnpikes had shown that the users were willing to pay rates varying between 1 and 1½ cents per mile for cars and lorries, depending on the size. The rates proposed for cars travelling on the proposed New Jersey Turnpike averaged 1½ cents per mile for the whole length, slightly more than 1 cent per mile at the southern end and in the more congested area of the northern end, 3½ cents per mile.

An attempt was then made to determine the percentage of potential vehicles travelling between adjoining States which might be attracted to the turnpike, traffic likely to be generated by the turnpike and of the future growth of traffic. The future plans of other States for highway construction and improvement were also considered in so far as they might affect traffic movement through New Jersey.

In the light of these facts, it was estimated that the annual increase would be 6 per cent per year up to 1956 and beyond that an arithmetical progression, namely 7 per cent in 1957, 8 per cent in 1958, etc.

The final analysis of economic feasibility indicated that if constructed and opened in 1952 it would barely earn its interest during that year, but revenue would grow quickly and the debt would be extinguished in about 25 years. It was decided, however, to go ahead with the project and finance it by the issue of $200 million tax free bonds.

A target period of 22 months was set from the receipt of the proceeds of the bond issue to the completion of the project. This was based on the work being carried out during two construction seasons with earthworks and bridge foundations in the first and paving and bridge superstructures in the second.

It had been determined that the turnpike should be constructed as a motorway with complete control of access. The design speed for the southerly 85 miles was 75 miles per hour and in the north, where more urban conditions prevailed, this was reduced to 60 and 70 miles per hour. Maximum gradient was to be 1 in 33 and the minimum radius of horizontal curves 3,000 feet. Hard shoulders ten feet wide and side verges six feet wide were to be provided adjacent to each of the carriageways. The width of the central reserve was to be 26 feet in the more rural areas in the south, where land was less valuable, reducing to 20 feet in the north; five feet on either side were to be stabilised as a safety device for vehicles travelling in the fast lanes.

The width of the traffic lanes was to be 12 feet and of the turnpike's 118 miles, 96 miles were to be constructed with dual two-lane carriageways. The remaining 22 miles were to be provided with dual three-lane carriageways. It was envisaged ultimately that, on certain lengths, the number of lanes would be increased from four to six and along the most heavily trafficked section the six lanes would be superseded by four carriageways each with two lanes.

The vertical clearance of bridges over the turnpike was limited to a minimum of 15 feet.

In formulating the design of the turnpike, it was found necessary, on a very few occasions, to resort to the minimum standards which had been laid down.

The first contract, covering earthworks and drainage, was advertised in mid-November 1949 and work started five weeks later.

In all a total of 80 major contracts were awarded, of which 55 were for over a million dollars each. The work involved the excavating of 52 million cubic yards of material, the building of 260 structures and the laying of 7 million square yards of paving in carriageways and hard shoulders.

The opening of the southern 93 miles in November 1951 after a construction period of only 23 months, constituted a record for this type of construction. This was even more remarkable when it is considered that the Authority had been formed only in March 1949, and only some 32 months before the opening of the first section.

The remaining section, which included the Passaic and Hackensack Bridges in the north, was completed in January 1952. It was then possible to travel a distance of 118 miles at a constant speed of 60 miles per hour from New York's George Washington Bridge south-west across New Jersey to the new Delaware Memorial Bridge in the State of Delaware; truly a worthy monument to the 700 engineers and 10,000 others who designed and built the turnpike.

The success of the turnpike is indicated by the increase in its use since it was opened. In 1952 the number of toll-paying vehicles numbered 18 million and, by 1965 nearly 65 million. The total revenue, including the income from concessions, increased during the same period from nearly $18 million to over $51 million.

In 1964 the Authority realised that consideration would have to be given to the possibility of widening certain sections at an early date.[1] Consulting engineers, appointed to carry out detailed studies of actual and future traffic, taking into account programmed highway development both in New Jersey and adjoining States, submitted their Report to the Authority in 1966. They concluded that along the southern part of the turnpike a decrease in future traffic could be expected arising from the construction of a parallel toll-free section of Interstate Route I–95 running in the same direction. In the north there was an entirely different picture. There the dual three-lane carriageways were carrying traffic densities ranging from 21 million to over 24 million trips per year, or an annual daily average of 63,000 to 66,000 trips in 1964 and 1965. Above the practical capacity of 80,000 trips per day congestion existed, accidents occurred and the turnpike suffered in respect of both safety and average travel speed. The number of days per year in which this figure had been exceeded had increased from three in 1960 to seventy nine in 1965.

It was recommended, therefore, that the improvements should consist basically of widening the northern quarter of the turnpike from the present dual three-lane carriageways to four three-lane carriageways, necessitating the reconstruction of several interchanges and the enlargement of existing service areas and maintenance depots. It was also proposed that a further length should be widened from the

[1] New Jersey Turnpike Authority, *Annual Report 1965*.

existing dual three-lane carriageways by adding two further two-lane carriageways.

The advantages of increasing the number of carriageways from two to four are many. Lorries and 'buses will, in general, be restricted to the two nearside carriageways whereas cars will have the choice of either carriageway. Having two carriageways in the same direction permits a flexibility which a single directional carriageway does not allow. In the event of an accident, emergency vehicles can reach the scene along the second carriageway without being delayed by vehicles stopped behind the accident. The cost of maintenance operations can be considerably reduced by diverting traffic during off-peak periods and thereby increasing the speed of the work.

The estimated cost of the improvements is approximately $300 million and the entire financial programme requires approximately $625 million in securities, including the retirement of all outstanding turnpike bonds.

Many lessons can be learned from the New Jersey Turnpike, of which the following are a few:

1. It has been clearly established that the provision of a toll facility of this type is an economic and acceptable proposition.
2. The extra efforts involved in bringing a new road into operation as quickly as possible after money has been made available for its construction are entirely justified. In this case, there was a need to produce revenue to pay off interest charges at the earliest possible date.
3. The successful completion of the design and construction in record time was made possible by the delegation of a considerable degree of responsibility to the Section Engineers within the framework of a sound policy.
4. The need to look ahead to future requirements is highly important. It is of interest to note that when sections of the dual three-lane highway became over-loaded the traffic studies indicated that additional capacity was best provided by conversion to four carriageways with a total of either ten or twelve traffic lanes.

Toll road financing in the U.S.A. reached its peak in 1954, when more than $2,000 million worth of bonds were sold. After that time the prospect of 90 per cent Federal Aid for the Interstate System had a considerable effect and virtually ended general interest in extensive further toll road construction except for projects outside the System.

A valuable contribution had been made, not only in the provision of many vital links in the national highway network, but also in the field of highway engineering.

Chapter 4

Justification and procedure for a motorway in Britain

Introduction

Before a motorway can be opened to traffic there are many legal, administrative and financial steps which need to be taken, as well as the detailed engineering design. The aim of this chapter is to summarise the guiding principles of the procedure, then to discuss the co-ordination and sequence of the various steps culminating in the building of the motorway. More detailed consideration is given in succeeding chapters to land acquisition and public relations, design and construction.

Guiding principles of procedure

A motorway, with its important influence on the social and economic well-being of an area, makes a significant contribution to the national economy and is constructed to benefit the community as a whole. The legal procedure has been laid down by various statutes for the purpose of protecting the rights of the private individuals and corporate bodies affected by the proposals, so that fair consideration may be given to their points of view. The Ministry of Transport is subject to the principle of public accountability and the present procedure involves administrative and financial processes which inevitably take up time and manpower.

The engineering aspect of the locating, designing and building of a motorway presents many complex and often unique problems. These have to be dealt with in logical sequence, from the first stage when growing traffic congestion and loss of life make the need for a motorway evident, through to the detailed supervision of its construction.

In general, the legal, land acquistion and other processes take much longer than the engineering design and, moreover, are subject to greater uncertainty because of the need to take into account the views of the interests affected, with the possibility that modification of the engineering proposals may be necessary. The Ministry is required by law to advertise the proposals at three main stages of the preparatory processes; first for the location stage, to establish the main line of the motorway, secondly for the alteration of side roads, public footpaths and private accesses affected by the route, and thirdly at the compulsory purchase stage for the land required.

Co-ordination of steps in procedure

The report *Efficiency in Road Construction* prepared in April 1966, under the auspices of the Economic Development Committee for Civil Engineering,[1] indicated that a major motorway project takes in the region of five years from the Minister's decision to proceed with the proposal, to the invitation of tenders, during which period for most projects the actual engineering design need not take more than two years. Recent developments may reduce this overall period of five years to less than three years. The *Second Report on Efficiency in Road Construction*, prepared in January 1967,[2] as a follow-up to the first Report, states that the Ministry of Transport are considering how to shorten completion of the legal and administrative procedure, including the fixing of the line of route and obtaining entry on to land. Much important preparatory work is required to justify the need for the motorway and to select the best route, before the Minister decides to proceed with the proposal.

The steps which have to be taken under the present procedure are inter-dependent and need to be carefully co-ordinated to ensure that each step is completed to a target date on a pre-planned programme. An overall picture of this co-ordination can be gained from Figure 8 which shows a network analysis for a typical motorway project. This technique is used to show how to do complex jobs in a logical way; the chain of arrows taking the longest time is the critical path which determines the total time taken for the whole job. It is important to have a completion date as the target to which everyone can work, knowing that their activity is contributing logically to this ultimate end.

[1] National Economic Development Office, *Efficiency in Road Construction*, London, Her Majesty's Stationery Office, 1966.

[2] National Economic Development Office, *Second Report on Efficiency in Road Construction*, London, Her Majesty's Stationery Office, 1967.

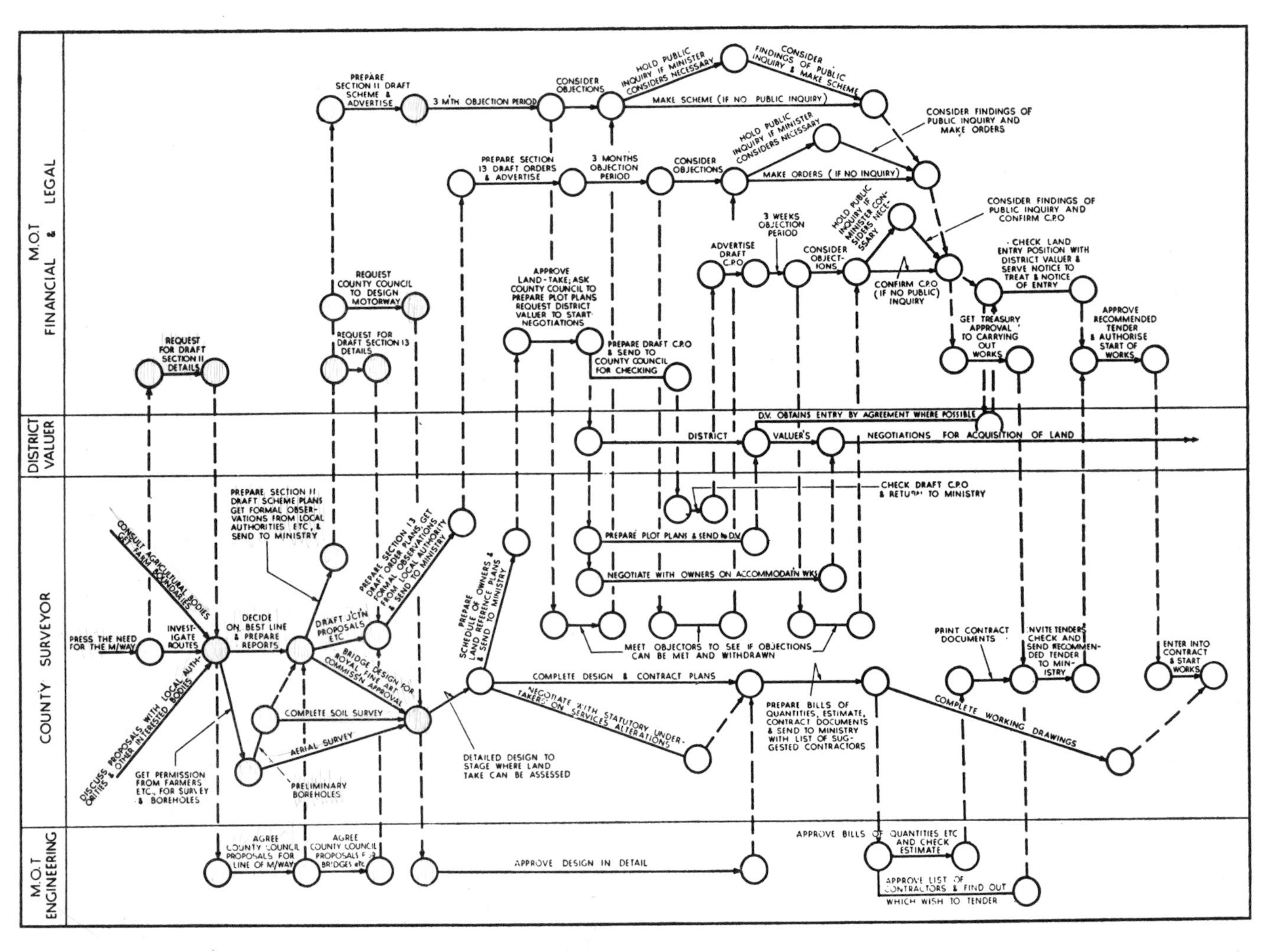

Fig. 8. Network analysis for a typical motorway project

Justifying the need for a motorway

The first necessity in setting the procedure in motion for the building of a motorway is to prove that it warrants a place in the Road Programme of the Ministry of Transport.

The Minister's report *Roads in England*[1] for the year ended 31st March 1966, stated that traffic volumes were expected to increase by one-third between 1966 and 1970 and to double between 1966 and 1980. The report admitted that even the present road programme could not hope to cope with the rising tide of traffic and the consequent spreading and intensifying of congestion. Competition for inclusion in the Ministry's Road Programme is very keen and to determine priorities a common yardstick is required to measure the respective merits of the various motorway proposals throughout the country.

For this purpose an economic assessment is prepared for the proposal, which may be defined as the rate of return to the community at large, resulting from the investment of the money required to build the motorway. This economic assessment and other factors which are taken into consideration in justifying the need for a motorway can be summarised as follows:

1. Will the motorway attract sufficient traffic, provide that traffic with a convenient route or take traffic off the most congested and dangerous roads?
2. What, in fact, are the conditions on the existing road system?
3. Will the road serve industry, developing areas or revive declining areas?
4. Will the road be a logical development in the regional and national motorway network?
5. Will it be reasonably cheap and easy to build?
6. In fact, will it be a good investment?

For a proposed motorway a very detailed investigation is carried out to measure benefits in relation to cost. The method, however, is still rather crude and is, in fact, frequently being refined by the Road Research Laboratory, whose latest modification[2] is so recent that there has been no opportunity yet of applying it.★ Despite these refinements, however, the fundamental principles remain and the method now described is that which has been used over the past seven years. This is based on a

[1] Ministry of Transport, *Roads in England for Year ended March 31, 1966*. Her Majesty's Stationery Office.

[2] Road Research Laboratory, *Research on Road Safety*, London 1960. Her Majesty's Stationery Office, and *Research on Road Traffic*, London 1965, Her Majesty's Stationery Office.

★ Authors' Note: This applied when this book was written. The Ministry of Transport's Technical Memorandum T5/67 is now followed for the calculation of economic assessments. This takes non-working time into account and gives an appreciable increase in the rate of return. The older method has been used in the examples quoted in this book.

paper entitled 'The Economic Assessment of Returns from Road Works'[1] presented in 1959 to the Institution of Civil Engineers by Dr. G. Charlesworth and J. L. Paisley, and subsequent modification by D. J. Reynolds[2] and R. F. F. Dawson,[3] both of the Road Research Laboratory.

The rate of return is simply the net annual benefit to traffic expressed as a percentage of the capital cost of the motorway. Benefits to traffic are derived from the fact that motorway travel is safer, faster and more efficient and therefore accidents, time, wear and tear, etc., all of which can be given a monetary value, are saved. In addition new traffic may be encouraged by the motorway. Other benefits such as improvements to amenity and reduction in human suffering, although real, cannot be valued.

The rate of return should be calculated for the traffic conditions at the time of opening of the motorway, which will generally be five years from the request for the justification. This means that the current traffic conditions must be projected forward five years. However, the monetary values assigned to the benefits and to the capital cost of the scheme need only be valued at the same year so it is usual to take these at the current rates.

Normally a proposed motorway route would need to provide an economic rate of return of at least 10 per cent to warrant consideration for inclusion in the Road Programme. As an example of a motorway in the Programme, for the proposed Manchester-Preston Motorway in Lancashire, the Chorley By-pass section would result in an economic rate of return of 64 per cent at 1970 level.

If a County Council is pressing for a motorway, the County Surveyor usually prepares the justification. If, however, the Ministry itself has considered a scheme to be worth investigation, it may instruct the justification to be prepared by the County Surveyor of the appropriate County, or a Consulting Engineer specially appointed for the purpose.

At this stage, all possible alternative routes must be considered impartially and evaluated. Where necessary, comparative traffic and economic assessments must be made, based on censuses carried out over a broad band of country between the proposed terminal points of the motorway, with a detailed study of the effect of the proposed motorway on other routes in the road network. For example, for the Manchester-Preston Motorway, M.61, of length about 22 miles between the East

[1] G. Charlesworth and J. L. Paisley, 'The Economic Assessment of Returns from Road Works', *Proc. Inst. Civ. Eng.* 1959 14, 229–54.

[2] D. J. Reynolds, *Road Research Technical Paper No. 48*. London 1960, Her Majesty's Stationery Office.

[3] R. F. F. Dawson, 'Vehicle Operating Costs in 1962', *Traffic Engineering and Control*, January 1963.

Lancashire Trunk Road A.580 at Worsley and the Preston By-pass at Walton-le-Dale, a total length of 60 miles of motorway was studied.

This investigation is, to my mind, the most important stage, for the requirements of traffic and the other various factors are often in conflict. After having obtained all the supporting evidence, it is vital that the correct decision is made as to which is the best route that will fulfil adequately and economically its role as part of the motorway network, and that will be generally acceptable.

Having decided which route to recommend, a report is often required giving the justification for this particular route. This should list the alternative routes and the reasons why they have been rejected. Approximate estimates of cost which should take into account any exceptional circumstances particular to each route are also needed.

It is difficult to give an accurate estimate of the overall time for this preliminary step of route selection. Often various routes have been thought of over a period of years. For example, on the Lancashire-Yorkshire Motorway route in Lancashire there are two parallel roads named Yorkshire Road in Prestwich, constructed in the 1930's, and between which a strip of land has been reserved for a route to Yorkshire (Figure 12). This strip of land, although for an all-purpose road instead of a motorway, and narrower than that now found to be necessary, stands as a testimony to the vision of planning at that time. What could not have been foreseen then was the phenomenal rate of growth of road freight and passenger traffic.

Will the motorway attract sufficient traffic, provide a convenient route and relieve congested roads?

The basic data for deciding this is obtained from an origin and destination survey. This can be carried out by roadside interviews, or by handing out prepaid postcards asking for the required information, or sometimes in city areas by home interviews.

From the collected data the proportion of traffic that would use the motorway has to be assessed. This is done by comparing the journey times on the existing roads with those on the motorway, were it built, to ascertain savings in journey times. Traffic likely to be generated by any proposed developments, such as New Towns, must also be taken into account. Then the layout of the motorway, whether dual two-lane or dual three-lane, and the type of junctions may be decided. Provided the preliminary line for the motorway has been well chosen the results of this investigation will merely confirm the need for the motorway and supply the statistical data for the justification and the detailed design.

What are the conditions on the existing road system?

The more congested and dangerous are the existing roads, the greater the justification for the proposed motorway. Examination of the traffic and accident data for the existing roads will show how bad these conditions are. For example, the amount of traffic a road can comfortably carry depends on its width, alignment, junctions, frontage development and so on. Within fairly wide ranges the capacities of roads have been defined by the Ministry of Transport. Knowing the capacities and present traffic volumes the amount of overloading on the existing roads can be calculated. Many of Lancashire's roads are already three or even four times overloaded. This congestion is reflected in the journey times and in delays at overloaded junctions, and in this form may be included in the calculation of the rate of return.

The dangerous conditions on the roads are illustrated by calculating the injury accident rate. This averages 2·75 injury-accidents per million vehicle-miles in Lancashire.

Safer travel

On average in Lancashire, motorway travel is five to six times safer than on the existing roads. From Police accident records and current traffic flows the accident rates on the existing road network are calculated. The amount of traffic likely to transfer to the motorway is determined from the origin and destination survey data and projected five years forward. The estimated number of accidents which would be caused by this traffic, if it had to remain on the existing roads, is compared with the estimate of accidents which would occur if the traffic travelled by motorway. The average cost of an accident in 1967 was estimated to be £750. The saving in accidents valued in this way gives the benefit due to safer travel.

Reduced journey times

Journey times on the all-purpose roads are affected in two main ways: by delays at junctions, and by low-running speeds between them.

Figure 9 shows the cost per mile for the average vehicle varying with speed. This is shown for 1962 prices, which are the latest available.[1] but I have also shown my estimate of the prices in 1966, with and without a value for non-working time.

The origin and destination survey gives the amount of traffic diverted to the proposed motorway from the existing roads. The motorway will benefit not only

[1] R. F. F. Dawson, 'Vehicle Operating Costs in 1962', *Traffic Engineering and Control*, January 1963.

the diverted traffic but also traffic on the old route which will be speeded up by the reduced congestion. The benefit to potential motorway traffic is calculated by comparing its costs if it travelled at the speeds and experienced the delays on the existing roads with its costs using the motorway at 50 miles per hour and no junction delay. The benefit to residual traffic is calculated by comparing its costs on the existing roads without the motorway with its costs at the increased speeds and reduced delays after the transfer of traffic to the motorway.

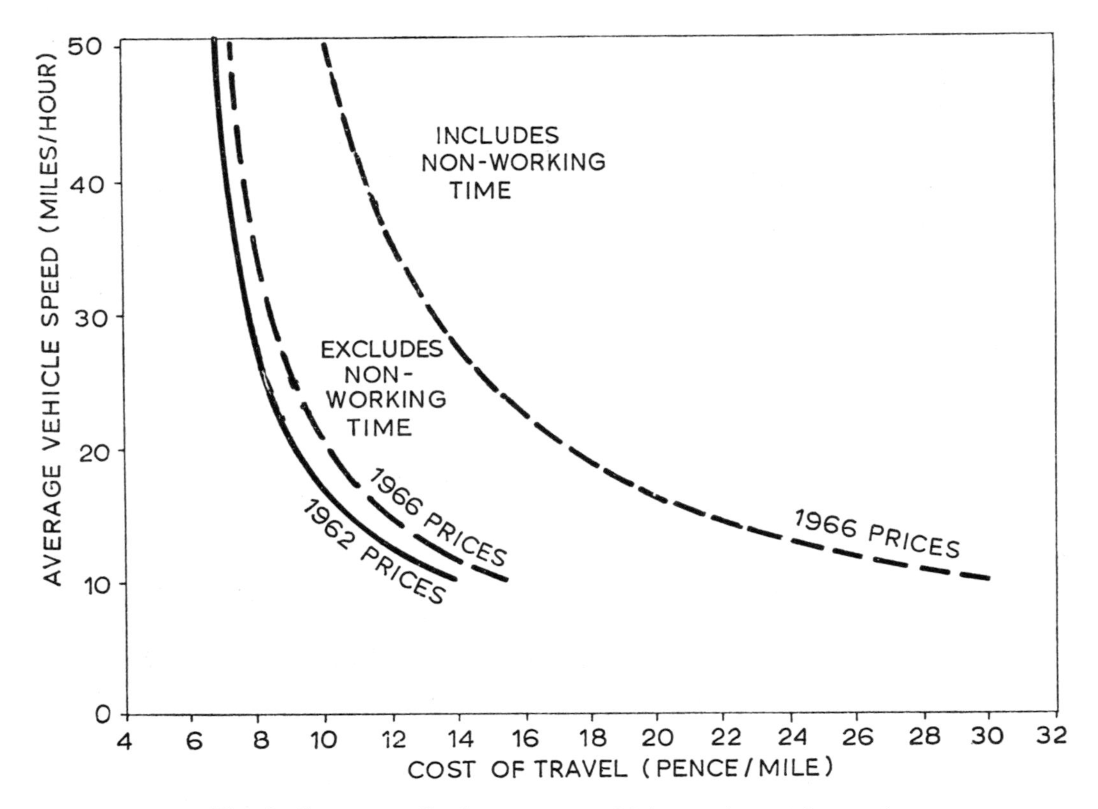

Fig. 9. Cost per mile for average vehicle varying with speed

In motorway design there is generally an allowance made for generated traffic. This is, to some extent, a safety factor and is necessary because it has been found in the U.S.A. and Britain that traffic actually using a motorway exceeds by 20 to 30 per cent the volume suggested by the O. and D. Survey. Some of this extra traffic will come from roads well outside the scope of the survey but much will result from completely new journeys, made worthwhile by the existence of the motorway. This traffic is arbitrarily valued at half the rate of normal transferred traffic.

Examples of economic assessments

The calculation of rate of return is illustrated by examples of two of Lancashire's proposed motorways: the South Lancashire Motorway and the Preston Northerly By-Pass. The earliest possible opening to traffic is 1972, but as the examples are taken from reports prepared a few years ago, they relate to 1970. In neither case in

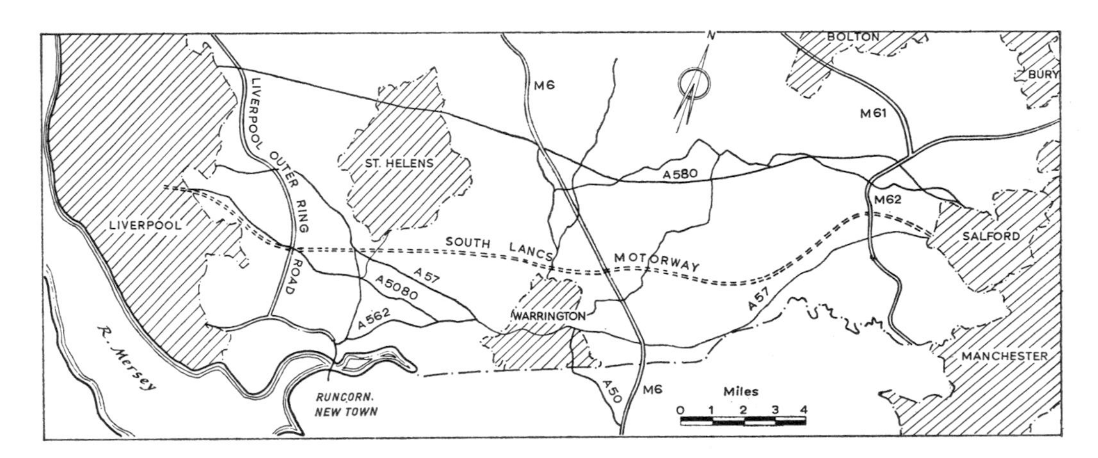

Fig. 10. South Lancashire Motorway

calculating savings in journey times has a value been placed on non-working time; were this done the rates would be roughly doubled.

Traffic which would benefit from the construction of the 25 miles of the South Lancashire Motorway between Liverpool and Salford now travels on the all-purpose road network shown in Figure 10. An origin and destination survey showed that the major transfer of traffic would be from A.580 and the A.57–A.5080–A.562 network, with smaller transfers from other roads. Allowance was also made for traffic likely to be generated by known new developments including Runcorn New Town.

Safer travel

The average personal injury-accident rate is 1·5 per million vehicle-miles on A.580 and 3·0 on A.57–A.5080–A.562. In 1970, without the motorway the expected number of injury accidents on these roads would be 720. With the motorway at an accident rate of only 0·4, 117 accidents would have been caused, saving 603 accidents. Valued at £750 each, this saving amounts to £452,250.

Reduced journey times

In 1964 the average vehicle speed between junctions was observed to be 41 miles per hour on A.580 and 30 miles per hour on A.57–A.5080–A.562 (20 miles per hour in the centre of Warrington).

Using the formula relating speed to volume and road width, these observed speeds are adjusted to August 1970 traffic conditions. There is no change on A.580 because it is a dual carriageway but the speed drops to 25 miles per hour on A.57–A.5080–A.562.

Potential motorway traffic would, in 1970, without the motorway, have travelled on A.580 at 41 miles per hour, or using the graph in Figure 9, 7·6*d.* per vehicle-mile, and on A.57–A.5080–A.562 at 25 miles per hour or 9·2*d.* per vehicle-mile, resulting in travel costs of £32,400 per day.

If it could have travelled on the motorway at 50 miles per hour or 7·2*d.* per vehicle-mile, the travel costs would have been £26,700 per day, saving £5,700 per day.

Traffic remaining on the old road system will benefit by an increased speed due to reduced traffic, except on A.580, a dual carriageway. On A.57–A.5080–A.562 average speeds will increase from 25 to 30 miles per hour, resulting in a saving of £1,600 per day.

At major junctions potential motorway traffic would experience delay valued at £1,800 per day and in the absence of intersections on the motorway all would be saved.

Traffic remaining on the old road system would experience a reduction in delay due to reduced congestion, amounting to £2,600 per day, while saving for vehicles at minor side roads is estimated at £700 per day.

Generated traffic

An allowance of 30 per cent is made for generated traffic over the traffic estimated from the origin and destination survey estimate as transferring to the motorway; and this benefit amounts to £900 per day. Total benefits due to reduced journey times thus amounts to £13,300 per day, equivalent to an annual benefit in 1970 of £3·99 million.

Maintenance costs

The extra maintenance cost due to 25 miles of motorway is £57,500. On the basis of these estimates, with the capital cost of the motorway, including land, estimated at £24 million, the rate of return works out at 18·3 per cent as follows:

Safer travel	£452,250
Reduced journey times	3,990,000
Total benefit	4,442,250
ddt. Extra Maintenance	57,500
Net benefit	4,384,750
Capital cost	24,000,000

Rate of return $\frac{\text{Net Benefit}}{\text{Capital Cost}} \times 100$ per cent $=$ 18·3 per cent.

Similar calculations for the $11\frac{1}{2}$ miles of the Preston Northerly By-pass (Figure 11)

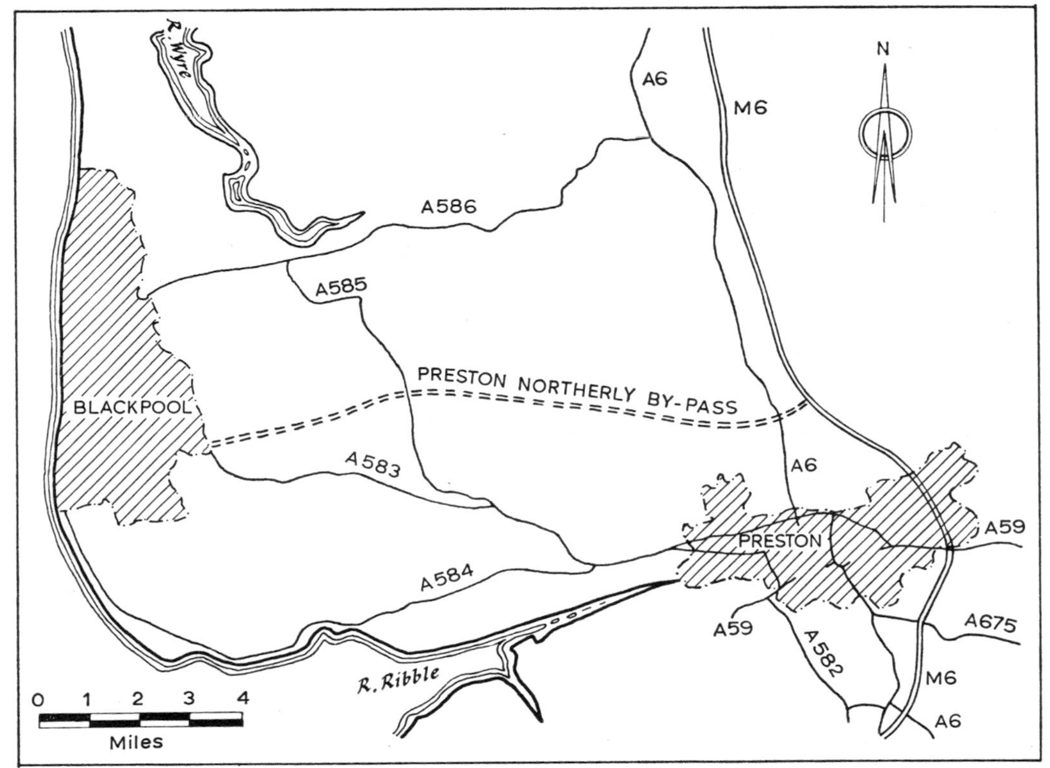

Fig. 11. Preston Northerly By-pass

between M.6 and Blackpool gave a 47 per cent return on a capital cost of £9·25 million as summarised below:

Safer travel	£383,250
Reduced journey times	3,990,000
Total benefit	4,373,250
ddt. Extra maintenance	26,400
Net benefit	4,346,850
Capital cost	9,250,000

Rate of Return $\frac{\text{Net Benefit}}{\text{Capital Cost}} \times 100$ per cent $=$ 47 per cent.

Selecting the route

If the Minister is convinced by the economic justification report and other relevant factors that the scheme warrants inclusion in the Ministry Road Programme, the most that can be hoped is that it will be included in the next roll-forward of the four-year programme. This is because a programme for the next four years will already have been settled, but this period is none too long as there are so many steps under the present procedure which must be taken before construction can start.

The next step is for the Minister to decide that the route for the motorway should be investigated. The invitation to undertake this is issued either to the appropriate County Council, on whose behalf the County Surveyor would carry out the investigation, or to a consulting engineer.

This investigation to select a route will, to some extent, overlap with the information obtained for the preparation of the justification for the route.

The investigation to select a route requires careful and detailed study of the requirements of traffic in order that the motorway may provide a practical and realistic solution to the problem of coping with this traffic, due regard being paid to the many factors which affect alignment. The influence of some of these factors is given below:

(a) ensure that the minimum disturbance is caused to existing and proposed land use, including agriculture, residential or industrial development, schools, public open spaces, etc.;
(b) avoid damage to and, if possible, enhance existing amenities, blending the motorway into its surroundings and paying due regard to aesthetic and landscaping considerations;
(c) take advantage of existing topographical features in order to provide a safe and flowing alignment, and keep cut-and-fill volumes to the minimum possible, consistent with meeting the standards prescribed by the Ministry for motorway design in rural and urban areas;
(d) have regard to the existing and future use of railways, canals, roads, footpaths and public utility services and determine which, if any, can be closed or diverted, so that the cost of providing crossings is kept to the minimum.

The appraisal of these factors involves informal consultation with Local Authorities including the Planning Department of the County Council, Statutory Undertakings, National Coal Board, British Railways, British Waterways, River Authorities, Ministry of Agriculture, Fisheries and Food, the Police, the National Farmers' Union and other bodies. Also, where necessary, liaison with adjoining County Councils and County Boroughs is required.

Particular note must be made of any burial grounds, common land, public open space or National Trust property which may be affected by a route, as these involve lengthy special Parliamentary or other procedure.

If possible it is preferable to avoid areas with special engineering difficulties, such as those liable to flooding, natural watercourses inadequate to receive motorway drainage, mining subsidence, extensive weak soils, peat, etc.

Fixing the line of the motorway

When the Ministry has agreed the recommended route, the necessary details are prepared by the engineer for advertising the line of the motorway. This is made under Section 11 of the Highways Act, 1959. The advertising of the line of the connecting roads at junctions with the motorway is sometimes incorporated in the same plan, which would then indicate both the centre line of the motorway and the connecting roads.

More detailed work is now necessary on the selected route and the basic layout at junctions and side roads established, so that the draft longitudinal profile along the motorway can be designed.

Initial thought must be given to the respective locations of a service area and maintenance compound for the motorway, so that the works in these areas can be planned and co-ordinated with the motorway construction, with both becoming operative at the same time as the motorway.

Informal consultations continue during this period with individuals, public bodies, the National Farmers' Union and other interested organisations. Arrangements are made for the route to be inspected by the Ministry's Advisory Committee for the Landscaping of Motorways and Trunk Roads.

Financial approval is obtained from the Ministry to the letting of a contract for detailed soil survey along the route and from the results of the bore-holes a report is prepared, on which will be based the engineering design and land acquisition.

Arrangements are also made for the route to be surveyed, usually by an aerial survey contract, so that large-scale contoured plans of existing features can be ready for the preparation of the detailed design.

The formal observations of Local Authorities and other appropriate bodies are obtained when the draft scheme plan, giving the proposed line of the motorway, has been prepared. These are forwarded to the Ministry with the draft Scheme plans of the motorway line and supporting information. Figure 12 shows part of a draft Scheme plan for a length of the line of the Lancashire-Yorkshire Motorway, M.62, in Whitefield and Prestwich, Lancashire. Figure 13 is part of the Scheme plan for the

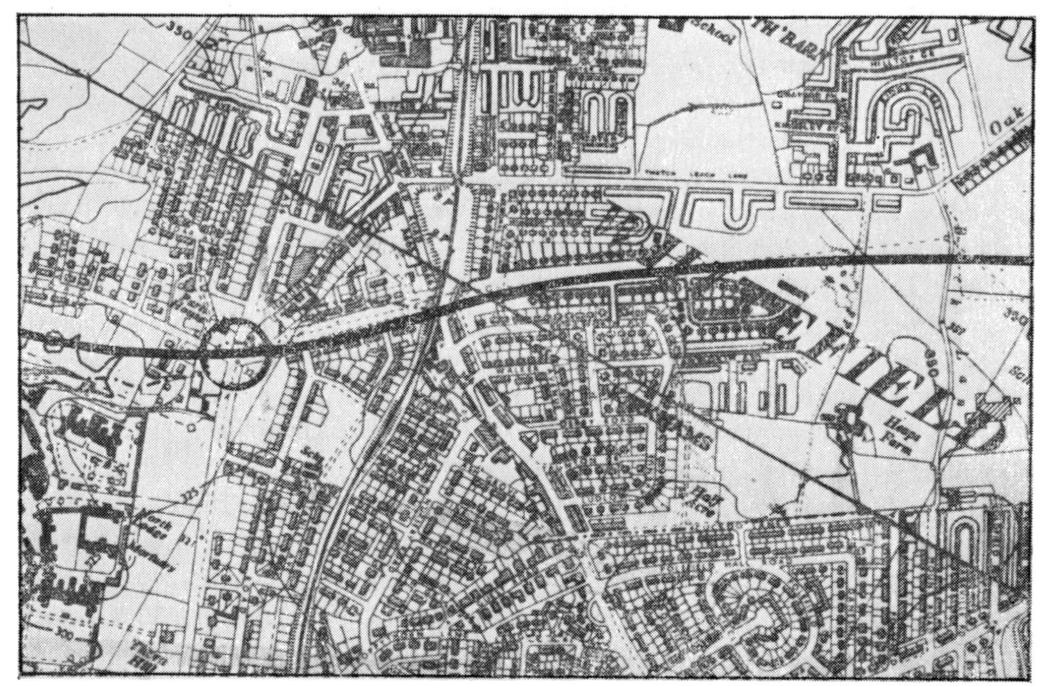

Fig. 12. Part of Draft Scheme Plan for Line of Motorway. Lancashire–Yorkshire Motorway, M.62

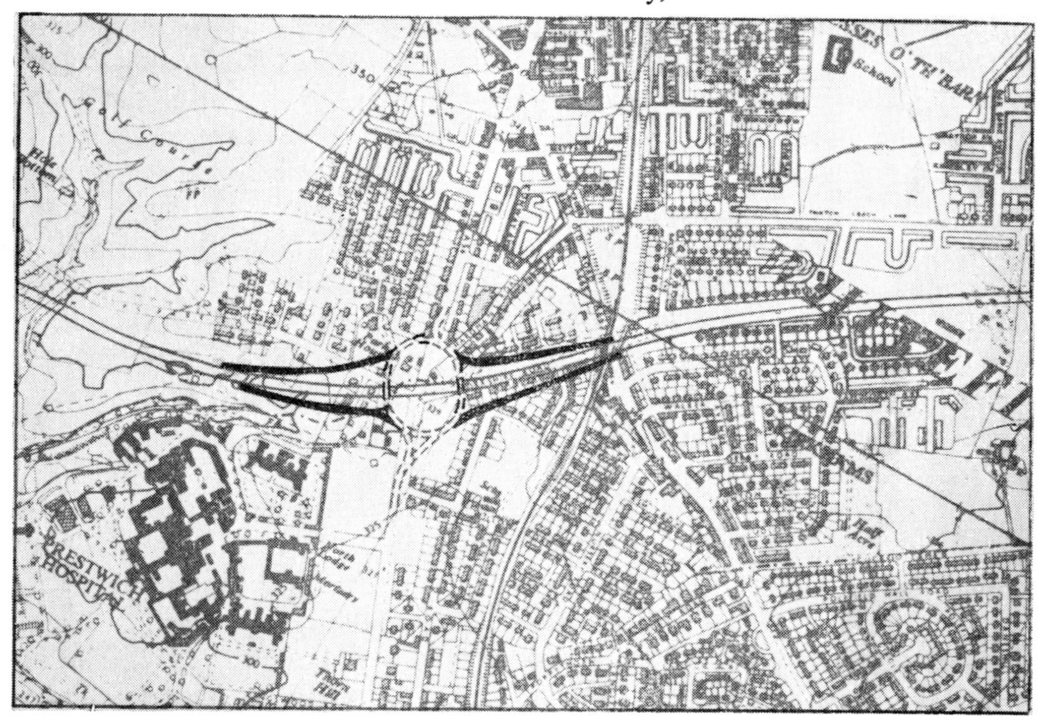

Fig. 13. Draft Scheme Plan for connecting roads at interchange between M.62 and Bury New Road, A.56

connecting roads at the interchange with Bury New Road, A.56. The relevant plans which follow in this chapter will be of this same area, so as to provide a better illustration of their purpose.

The Ministry now prepares the draft Scheme for fixing the motorway line from the information provided by the engineer. The observations of the Regional Land Commissioners, Ministry of Housing and Local Government, the Licensing Authority and Traffic Commissioners are obtained by the Ministry of Transport, who then advertise the Scheme. Three months is allowed for any person or body to object whether or not his property is affected.

All objections are given detailed consideration. If required, a meeting with objectors is arranged jointly by the Ministry and the engineer to explain the reason for the proposals and to see if the points of the objection can be met. The Ministry writes to the objectors, giving the relevant explanations, and asks if they are prepared to withdraw their objections, or wish them to stand.

The Minister must then decide whether the weight of objections justifies the holding of a Public Inquiry or whether to confirm the line of the motorway, with or without modification. If an objection is received from a Local Authority a Public Inquiry must be held. At this stage thorough investigation in choosing the selected route pays dividends, because if the choice is sound, and the public is kept informed of the reasons for this choice, there is far less likelihood of a Public Inquiry being needed, or of any substantial objection.

If, after consulting the Inspector's Report on the Public Inquiry, the Minister decides to abandon the route, it is necessary to go back to the beginning and advertise another route. In Lancashire, we are fortunate that with 67 miles of motorway open to traffic and a further 40 miles of motorway line fixed, this eventuality has not arisen. This is due in great measure to the helpful co-operation received from the public and the Local Authorities and other bodies affected.

After publication of the confirmation of the line of the motorway by the Ministry, there is a further six weeks' period during which objections can be made on the legality of the scheme, that is, that the statutory procedure has not been followed fully.

Alteration of side roads, footpaths and private accesses

At an appropriate stage after the invitation is issued by the Ministry to the County Council to prepare the Scheme for the motorway line, a similar invitation is issued for the preparation of Orders for alteration of side roads, footpaths and private accesses, together with the preparation of land plans and contract details. These alterations generally are covered under Section 13 of the Highways Act, 1959, but if a trunk road is involved, separate Orders are required under Section 7 or 9 of this Act.

The alterations are designed in sufficient detail to prepare the draft Order plans. Here again, informal consultation is required with individuals, public and other bodies affected, before the proposals can be formulated.

With a similar procedure to that for the main line of the motorway, the completed draft plans for the alterations to side roads, etc. are sent to Local Authorities, Statutory Undertakings and other bodies for their observations. These are then sent to the Ministry with the draft Order plans (Figures 14 and 15) and relevant details.

The Ministry prepares the draft Orders for such alterations from the information provided, in a similar way to the Scheme for the motorway, and advertise the Order.

Three months are allowed for any person or body to object and these are considered as for the Scheme.

Arrangements have been made with the Ministry on more recent motorway routes for the side road Orders to be advertised during the period of objection to the Scheme for the motorway line. This has the advantage of minimising the time taken in legal procedure and is more helpful to the public, because it is often found in practice that points of detail are raised regarding farm and property access when the main Scheme is advertised. It is far better to be able to explain what is proposed at the time, rather than several months later. There is, of course, the risk that some of the design on the side roads may be abortive, if there is any substantial amendment to the motorway line.

Purchase of land

The procedure in the acquisition of land for a motorway is referred to in more detail in Chapter 5.

After the preparation of the side road alterations previously described, the engineering design is completed by the engineer in sufficient detail to prepare land reference plans and schedules and these are sent to the Ministry so that the draft Compulsory Purchase Order can be advertised. Plot plans for the individual land interests are also prepared, ready for use by the District Valuer in his negotiations. Examples of land plans are included in Chapter 5.

When the Minister advertises the draft Compulsory Purchase Order, objections may be made by the property interests directly affected during a three-week period. The procedure concerning consideration of any objection is similar to that for the main line of the motorway. A Compulsory Purchase Order cannot be confirmed

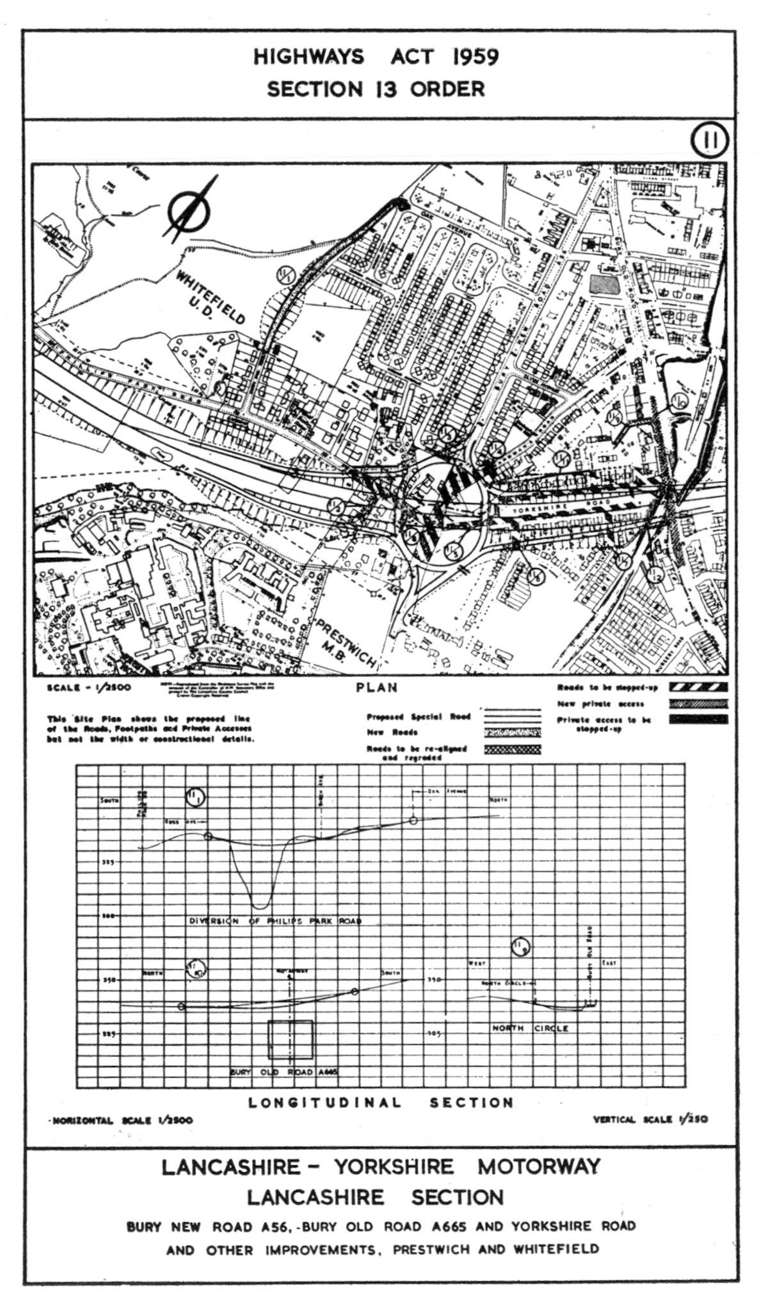

Fig. 14. Draft Order Plan for side road alterations, etc., near Bury New Road, A.56 for M.62

until the relevant Scheme for the motorway line and Orders for the side road alterations have been confirmed. When the Compulsory Purchase Order is confirmed and

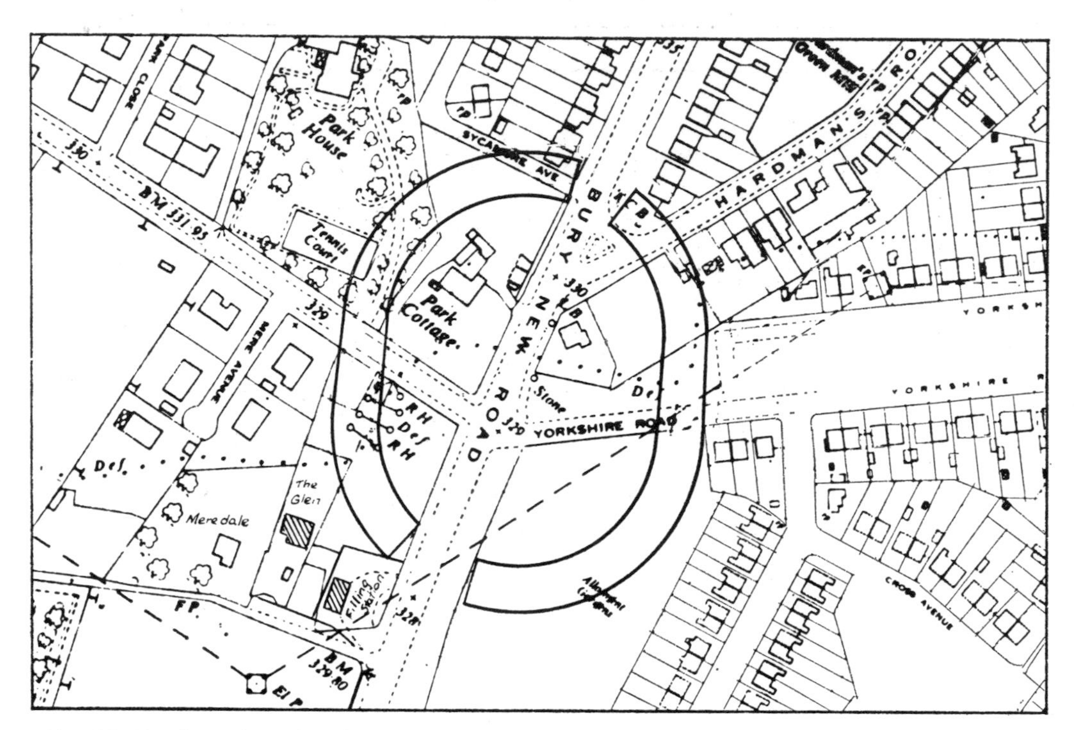

Fig. 15. Draft Order Plan for trunk road alterations to Bury New Road, A.56 for M.62

entry will be legally possible on all the required land, tenders can be invited for the construction of the motorway.

Design and contract documents

Ideally, the completion of printing of the Contract Documents should coincide with the date of clearance of the land-entry position. In order to arrive at this stage, the detailed road and bridge design must be completed and Contract Drawings prepared (Figure 16). For this purpose, selective use of computers is made to increase output and cut down staff time spent on tedious, repetitive calculation. For the bridges the designs may have to be submitted to the Royal Fine Art Commission for approval (Figure 17). Detailed consultations are held with the Ministry's Landscape Architect and Planning Department of the Local Authority on landscaping proposals.

The Contract Drawings are then used to prepare the Bill of Quantities which is

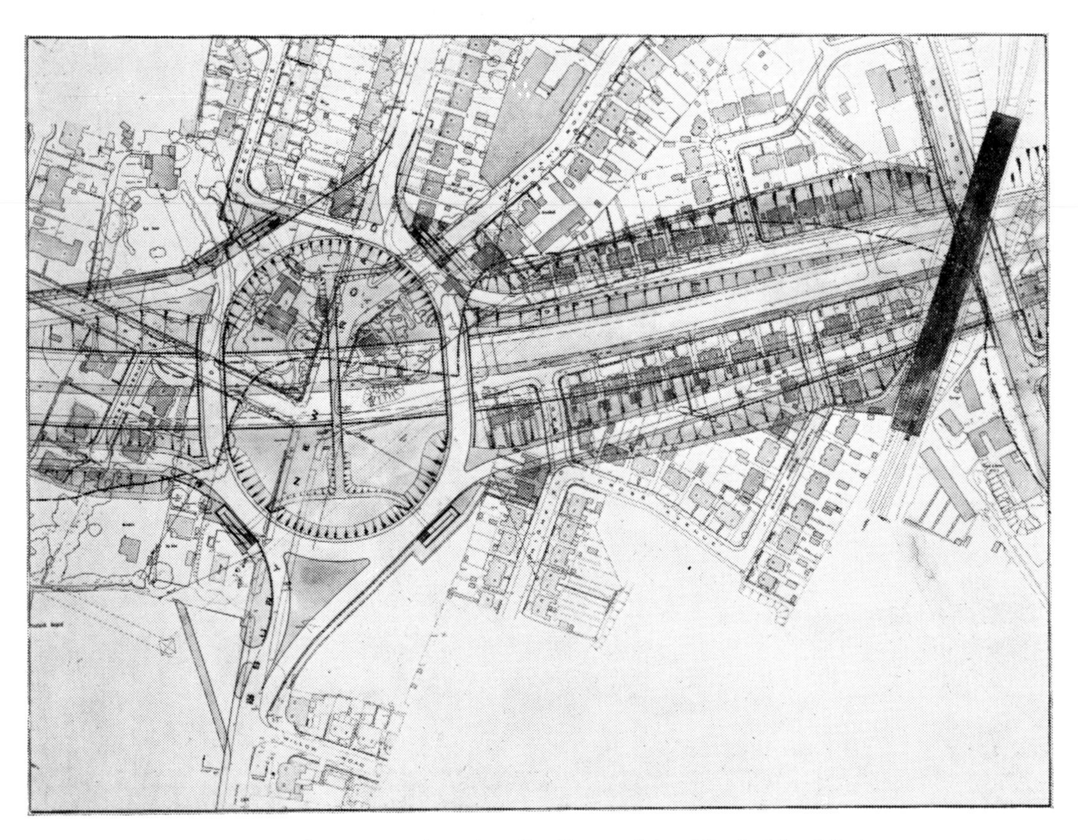

Fig. 16. Part of contract drawing for Lancashire–Yorkshire Motorway

Fig. 17. Snowhill Lane bridge on M.6. Perspective sketch

incorporated in the Contract Documents, from which an up-to-date estimate of cost is prepared. The documents, drawings and estimate are sent to the Ministry for engineering, contractual and financial approval. When this has been received, the Contract Documents can be sent for printing.

Table 5
Number of drawings prepared for M.6 in Lancashire

Description	Negatives
General Plan, Setting Out Plan and Drainage Plan	250
Junctions, Side Roads and Footpaths	450
Motorway Cross-Sections, Miscellaneous Typical Details	350
Land Reference Plans, Plot Plans, Accommodation Works	5,000
Bridgeworks	3,000
Soil Profiles, Mains and Services, Culverts, Outfalls, etc.	450
Site Drawings	500
Total	10,000

The extent of the drawing and design work involved may be gained from Table 5 listing the number of drawings prepared for the M.6 in Lancashire. From each of these negatives, printed copies were obtained, the number depending upon their particular use.

Tenders can now be invited. This is usually by selective tendering and the list of contractors must previously be cleared with the Ministry, so that there will be no delay in sending out the Contract Documents and Drawings.

A tender period of ten weeks is usual for a major motorway project and when the tenders are received, they are checked and agreement of the Ministry is obtained to the successful tender.

The final steps

Immediately prior to awarding the Contract, the organisation for supervision of the Contract by a Resident Engineer and his staff is set up. Actual construction could take from eighteen months to three years. Even after the completion of construction, the engineer's work is not complete, for there is still the settlement of final accounts with the contractor.

The final step in the procedure, when the motorway works are nearing completion, is the advertising by the Minister, in the *London Gazette* and at least one local paper, of the date on which the route will be open for use and on which the various motorway regulations will come into operation. The Notice is made under the provision of the Special Roads (Notice of Opening) Regulations, 1962, published as Statutory Instrument 1962/1320.

Summary

There have been many criticisms of the existing statutory procedure. The problem is to reconcile speed with adequate safeguards for the preservation of democratic rights, but there is, in my opinion, no doubt that some streamlining can be effected without prejudicing the latter.

On the one hand lives are being lost unnecessarily on the congested all-purpose roads which the motorway will relieve. On the other hand one must ensure that a fair hearing is given to those affected by the route, and keep any hardship to the minimum.

Productivity in motorway design and construction presents a continual challenge to ensure that each motorway contributes to the national economy in the shortest possible time. The organisation of the stages in the programme, therefore, is one of constant re-appraisal of priorities, to ensure that no avoidable hold-up occurs and that abortive work is kept to the minimum.

The backroom work is enormous and complex. This chapter has been included to give some idea of what is needed before a new motorway route can be opened to traffic and help to make the areas near the old congested routes safer and more pleasant places in which to live.

Chapter 5

Land acquisition and public relations

The approach to the people affected

Acquisition in this context means the purchase of land by a Highway Authority resulting from negotiations on purchase price between the District Valuer and the members of the public whose properties are affected by the proposals. The District Valuer is responsible for recommending to the Highway Authority that the agreed purchase price is fair, in terms of the compensation law applicable.

The public relations aspect is vital in connection with land acquisition and it is essential that a good public relations basis is established well before negotiations for acquisition are opened.

To give a clear understanding of what is involved it is necessary to start at the point when a member of the engineer's staff has the first all-important meeting with the person who either lives, or has an interest, in land on the line of the motorway. In the County Surveyor's Department in Lancashire there is a separate Land Section and its members often make the first contact with the landowner and tenant. Because of their training, this first meeting is usually mutually beneficial. It takes place when the necessity for a motorway has been established and a practical and acceptable route found. Its ultimate feasibility can only be proved by going out on site to survey existing features, take levels and carry out the soils survey. This inevitably means entering upon land owned, leased or tenanted by a member of the public who has a legal right to be there. Up to this time, the owner or tenant has not become involved in the project, although he may have heard rumours, or seen vague reference to the motorway proposal in the local Press. However, the land surveyor representing the engineer can explain personally better than by letter what it is hoped to achieve and how, and what rights the individual has. He can often allay any fears or concern which may exist–and they do exist because houses are homes and business premises and farmlands are a means of making a living. The Authority also benefits because, having established the foundation of an amicable relationship, the proposal can proceed with maximum efficiency, with due regard paid to the individual's rights.

Permission to enter the land for survey purposes is the immediate aim of this first meeting. The land surveyor will explain that compensation will be paid for any damage done to the land, which will be kept to the absolute minimum and, if possible, avoided completely.

Land plans and schedules

Arising out of this work, and assuming all the investigations prove that the motorway is a practical proposition, the Scheme is published advertising the centre line of the motorway. Afterwards a further Order is advertised for the diversion and/or closing of all side roads, public footpaths and private access roads which are intersected by the proposed motorway. Advertising of the Scheme, and Orders which provide the necessary legal powers to negotiate to acquire the land, give the public the first opportunity to object officially.

Next, the detailed engineering design of the motorway–road, bridges, drainage, etc.–is completed, and from this a plan is produced in colour showing exactly which land is required. It can now be said we know exactly what we want but we do not know from whom we have to buy it.

It now becomes the job of the engineer's land surveyor, furnished with this coloured plan, to obtain full details of the legal interests of every square yard of land needed for the motorway. A person's legal interest may be as owner, lessee or tenant and the precise nature of this interest must be found. This information is obtained by meetings with the individuals concerned and their solicitors, and from inspection of deeds.

Based upon the information obtained, the following are now produced:

(i) Land reference plans;
(ii) Land reference schedules;
(iii) Interest or plot plans.

(i) Land reference plans (see Figure 18) show the various legal rights which the Authority wishes to acquire, viz. land to be purchased and other rights such as easements, which are necessary for drainage requirements and bridgeworks–in fact anything essential for the construction of the motorway. All boundaries of legal interests in the land required are shown and each legal interest allotted a plot number.

(ii) The land reference schedule (see Figure 19) is read in conjunction with the

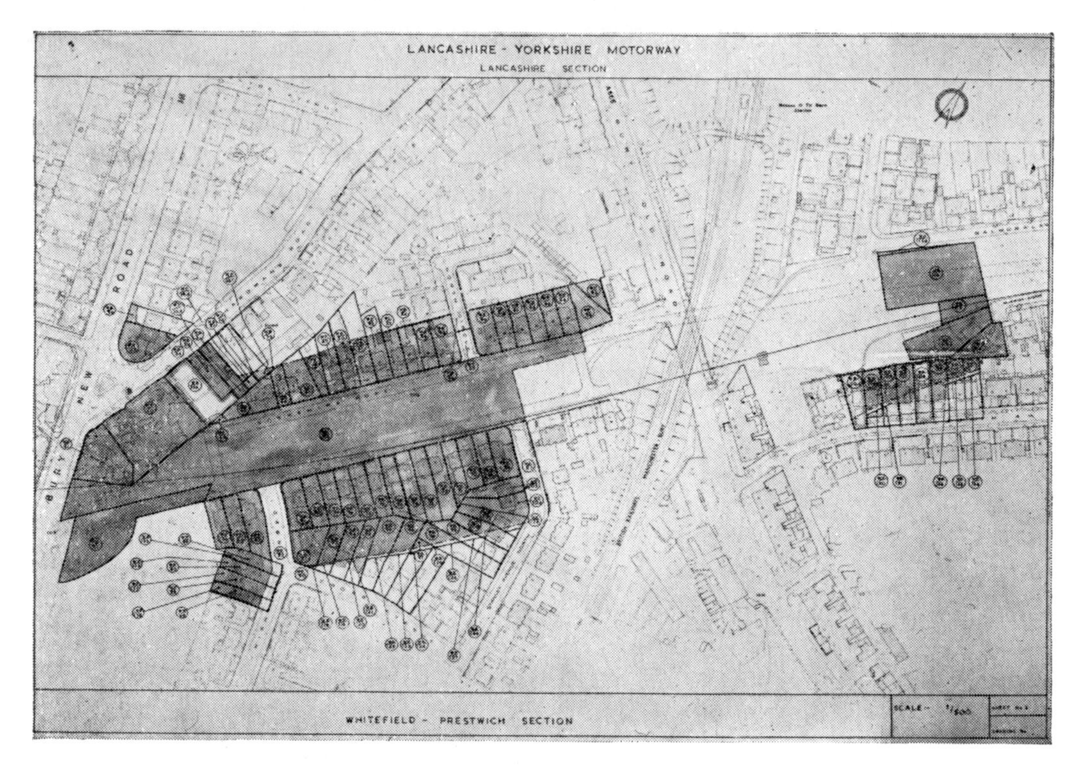

Fig. 18. Land reference plan

Name of Road: LANCASHIRE - YORKSHIRE MOTORWAY (Lancashire Section)
Scheme: WHITEFIELD - PRESTWICH SECTION

COUNTY OF LANCASTER

SCHEDULE OF LANDS

Date

(1)	(2)	Area (3)	Description and Situation of Land (4)	Owners (5)	Lessees (6)	Occupiers (7)	Remarks (8)
TITLE	WP 74	371 sq. yds.	In the Parish of Prestwich: Semi-detached brick-built house with tiled roof, nod. 29, Yorkshire Road, Prestwich, together with timber and asbestos garage, outbuilding and gardens to the front and rear	Trustees of James Smith Dec'd. 1. Herbert Smith, 16, Main Avenue, Manchester 2 George Jones, 2, Manchester Road, London E.C.2	Joint Lessees. Bernard Brown, and his wife Mrs. May Brown, 29, Yorkshire Road, Prestwich, Manchester.	Lessees.	Mr & Mrs. Brown hold the land on a lease for 999 years commencing 25th March 1930 at a ground rent of £6 p.a.
TITLE	WP 75	317 sq. yds.	Semi-detached brick-built house with tiled roof, nod. 27, Yorkshire Road, Prestwich, together with outbuilding and gardens to the front and rear.	Alfred Morgan, 27, Yorkshire Road, Prestwich, Manchester.	—	Owner.	
TITLE	WP 76	285 sq. yds.	Semi-detached brick-built house with tiled roof, nod. 25, Yorkshire Road, Prestwich, together with timber and asbestos garage and gardens to the front and rear.	Drinkwell Brewery Ltd. Reg. office:- The Brewery, Hops Road, Manchester.	—	Brian Walker, 25, Yorkshire Road, Prestwich, Manchester.	Mr. Walker occupies the property on a service tenancy.
TITLE	WP 77	282 sq. yds.	Semi-detached brick-built house with tiled roof nod. 23, Yorkshire Road, Prestwich, together with timber and asbestos garage and gardens to the front and rear.	Property Investment Trust Ltd. Reg. office:- 1/3, High Street, London W.1.	H & B. Withnell (Builders) Ltd. Reg. office:- 10, Market Place, Rochdale. Sub Lessee:- Fred Albert, 23, Yorkshire Road, Prestwich, Manchester.	Sub Lessee.	H & B. Withnell (Builders) Ltd. hold the land on a lease for 999 eaRS commencing 1st June 1928 at a ground rent of £6 p.a. Mr. Albert holds the land on an under lease dated 1st December 1938 for the residue of the term of the head lease less 10 days paying a ground rent of £7 p.a.

Note. All names in cols. 5, 6, & 7 and information in col. 8 are fictitious.

Fig. 19. Land reference schedule

reference plan and virtually is a cross reference to it, providing names and addresses of all having a legal interest in the land, the type of interest (freehold or leasehold), a description of the land, its location and area. Details of the terms of any leases are also given.

(iii) The plot or interest plans (see Figure 20) are, in effect, a breakdown of the land reference plan, one being produced for each of the legal interests involved.

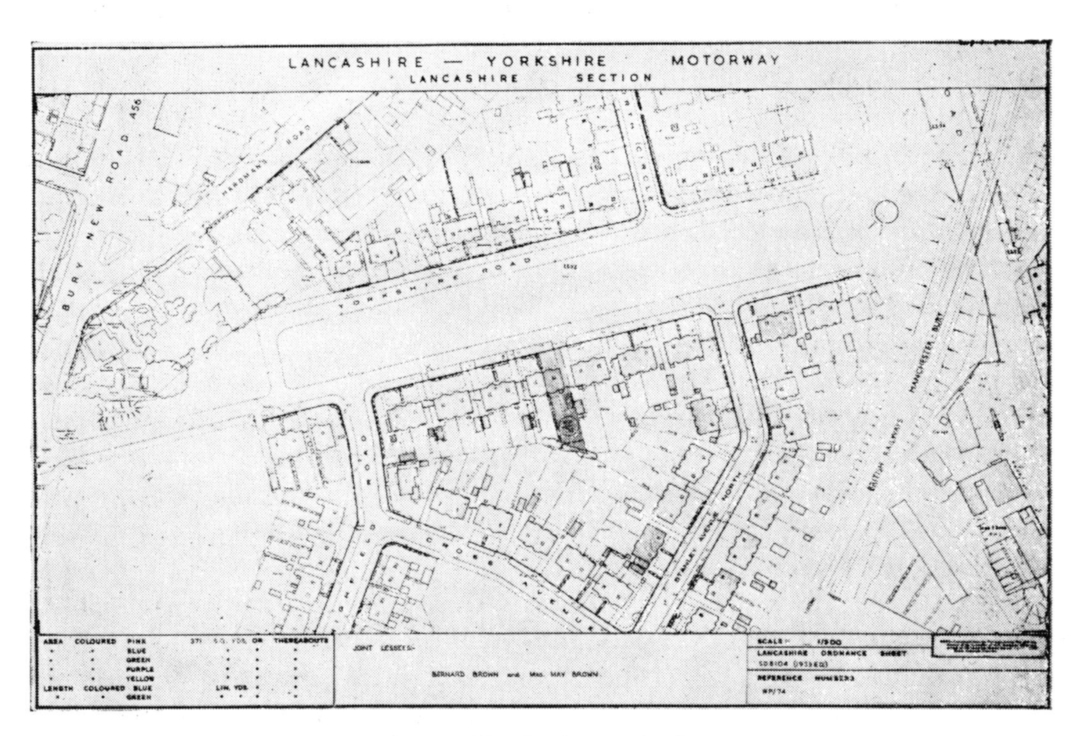

Fig. 20. Plot (or interest) plan

These plans indicate the motorway requirements as restricted to one person's interest, and include the name of the individual, his interest and the area of land illustrated.

Copies of the reference and interest plans and the schedule are forwarded to the Ministry of Transport and to the District Valuer.

Negotiations for land acquisition

Acquisition of land in the accepted sense can now begin. The District Valuer's office is a branch of the Inland Revenue, and the District Valuer and his staff conduct, on behalf of the Ministry of Transport, the negotiations relating to compensation.

Using the plans and schedules he contacts all the parties named, when instructed by the Ministry, asking them to open negotiations with him. Any member of the public whose land interest is directly affected is entitled to the services of a Valuer and Solicitor of his own choice during the negotiations and transactions which follow, and proper legal and valuer's fees are reimbursed by the Ministry of Transport. Compensation law is rather complex, and it is easy to take some comment regarding compensation out of context, apply it to a case to which it should not be applied, and feel that one has been misled or misinformed.

In opening negotiations, the District Valuer also states the target entry date upon which it is provisionally aimed to start construction.

During the same period, negotiations take place regarding accommodation works to be carried out on the land retained in private ownership for the benefit of the owner and/or occupier concerned. When completed satisfactorily, they are passed over to that owner as his, so far as future maintenance is concerned. A simple type of accommodation work is the rebuilding of a garden wall on an improvement line. However, on farms severed by a motorway, these works can be quite complicated and generally consist of all work necessary to ensure that the farm continues to be operative as a viable unit. This is not always possible and the loss of facilities of a property has to be taken into account on a pure compensation basis. Accommodation works on farms can consist of such things as provision of water supplies to severed land, gates, removal of hedges and fences to make more easily worked fields, construction of occupation roads, etc.

The agreed accommodation works form part of the negotiations regarding compensation: one is not necessarily set off against the other, but in total represent what is paid and done in exchange for the land acquired.

Copies of the land reference plans and schedules passed on to the Ministry of Transport by the engineer are used to record the acquisitions and as the basis for the Compulsory Purchase Order.

The number of legal interests affected

By its very nature a motorway stretching across the country has a significant effect on the property it passes through. The number of legal interests involved can vary appreciably, dependent upon the proportion of urban and rural land through which the route passes. On the 61 miles of M.6 now built in Lancashire, predominantly through rural areas, negotiations involved 737 people with a legal interest in land to be acquired; but 42 of these were confined to a one-eighth of a mile length of the motorway in the urban area of Orrell, near Wigan.

Compulsory purchase

When a Compulsory Purchase Order is prepared it is advertised, in the first instance, in draft form and three weeks are allowed for the submission of objections. If none is received, the Order can be confirmed; but any objection automatically precludes entry on to the land until it has been fully considered. It is usual to seek withdrawal of the objection by explaining to the objectors what is involved but, if this fails, then a Public Inquiry must be held.

The Inquiry is presided over by an Inspector who hears the authority's case and the reasons for the objection as provided by the person concerned. The Inspector submits his report to the Minister of Transport who may either confirm the Order in the draft form advertised, modify or quash it. The confirmation of the Order is advertised and six weeks is allowed for objections, but at this stage objections are restricted to points of law (applicable to the legal and statutory processes involved in the making of the Compulsory Purchase Order) and cannot be a repetition of the objection raised at the draft stage. Disagreement regarding compensation is not a valid objection to a Compulsory Purchase Order. Any differences of opinion on the District Valuer's valuation is resolved by the Lands Tribunal.

After confirmation of the Order the Ministry of Transport may serve on all parties affected a Notice to Enter, and after the expiration of fourteen days entry on the land can be made compulsorily.

Compulsory purchase is often considered by the public to be Government wielding a big stick. This is wrong. If we accept that the construction of a motorway benefits the community as a whole and that the Government is mindful of the rights of the individual affected, then Compulsory Purchase serves a double purpose. It provides an opportunity for the individual to object to the project and is an insurance against possible waste of public money.

Although a Compulsory Purchase Order is made for practically every scheme, it is not necessarily intended that it should be used. The time it takes to prepare, advertise and confirm a Compulsory Purchase Order depends on the size of a scheme, and also whether it is necessary to hold a Public Inquiry, but twelve months can be regarded as average. Entry by compulsory powers is effected only if all else has failed.

The Lancashire County Council have a good record for carrying out schemes by agreement and avoiding use of compulsory powers. The basic public relations approach has contributed to this. It is essential that the Engineer should be willing to meet people individually or collectively. By collectively I mean groups of people with a common interest or involvement who get together and form a Committee, or organisations like the National Farmers' Union. Generally speaking, members of the public are very willing to co-operate and objections stem more from lack of knowledge and fear of the unknown rather than the actual effect on their properties. By applying patience and understanding, showing a willingness to explain what is involved and to modify the proposals where justified, objections can often be averted or withdrawn. Acquisition of land does not involve cases but people.

The broad principles of compensation

The true basis of compensation is market value–that is the price agreed in a transaction involving the sale of land or buildings, between a willing seller and a willing buyer. Compensation must be limited to the extent of the legal interest held–for instance an owner/occupier is naturally entitled to more than the owner of a house which is tenanted, and similarly a person living in premises on, say, a seven-year lease must be considered on a different basis.

Compensation becomes more complicated when only part of an ownership is required. One then has to consider what is known as injurious affection which, in effect, is the degree related to money by which the value of land not being acquired is depreciated by the acquisition and use of the land required. This can probably be more readily understood by considering the difference between the 'before and after value' of the land interest. For example, in the case of a modern semi-detached house where, say, three-quarters of the depth of front garden is required for roadworks, the compensation must be more than the pure value of so many square yards of land being part of the front garden. Again, were a 50-acre farm severed by a motorway into two approximately equal areas, the motorway taking up say five acres of the farm, the compensation is appreciably more than the value of the five acres required for motorway construction on the edge of the farm.

Certain facilities have been introduced into the law over the years to eliminate hardship or inconvenience which would otherwise exist. These include the purchase of houses in advance of requirement following service upon and acceptance by the Authority of a Purchase Notice (Town and Country Planning Act, 1962, Section 130). This is operated when a person wishes to sell his house but is unable to do so because of an impending highway project which involves acquisition.

Hardship or inconvenience can also be alleviated by a provision which enables the person affected to obtain very quickly up to 90 per cent of a provisionally agreed amount of compensation should he so wish. Normally, as with private transactions, the money is paid upon legal completion which, in the case of a motorway, would usually not be until after the motorway contract has started or even after it has been opened to traffic. The 90 per cent provision is particularly necessary for someone buying a new house.

Interest on the compensation ultimately agreed is payable from the date of entry upon the land, so completion after the motorway has been opened does not necessarily indicate laxity on the part of the Authority. Quite often it is sensible for the person affected to defer the final agreement to compensation until the effect of the motorway can be seen, and therefore, more accurately assessed rather than assumed. A third facility which can alleviate possible hardship is that compensation is not necessarily restricted to the existing use value of the land. If the owner can obtain from the Planning Authority a certificate (Town and County Planning Act, 1962, Section 17) to the effect that, but for the construction of the motorway, it would have been permissible to build on the land, then building and not agricultural land prices will be paid.

The Ministry of Transport and the County Council are not Housing Authorities, but by liaison with the local Housing Authority, and payment of subsidies, it can be taken as a general principle that all tenants whose houses are required are found alternative accommodation. In certain instances, owner/occupiers may also be eligible for re-housing but these generally are confined to 'hardship', usually where age or infirmity is evident.

During construction

After tenders have been invited and the Contract awarded, the entry date on to the land becomes a reality and the successful contractor arrives on the site, sets out the motorway and erects temporary fences on the acquisition limits. So actual construction starts and throughout the period of the Contract, full liaison and co-operation with the public is maintained–on the one hand reducing to an absolute minimum any disturbance of the land and inconvenience to the people alongside the works, and on the other ensuring that the job proceeds smoothly.

Acquisition for county motorways

The procedure already outlined is applicable specifically to Ministry of Transport trunk motorways where the full cost of acquiring the land and building the road is paid by the Ministry. The title of the land is acquired in the name of the Ministry of Transport and the compensation is paid by the Ministry.

There are cases, such as the existing Stretford-Eccles length of M.62 in Lancashire, which have been built as county motorways with a grant from the Ministry according to the classification of the road. The procedures are very similar to trunk motorways as far as members of the public are concerned but, for county motorways, compensation is paid by the county, which also acquires the land.

Chapter 6

Motorway design

Introduction

The aim in designing a motorway is to convey the expected traffic safely and smoothly along a route which is economical to construct and maintain, with the minimum interference with land use.

In this Chapter, I will consider first the requirements of traffic, then the standards by which the motorway is provided to take this traffic. This will be followed by the main principles by which these standards are applied, so that the motorway is tailor-made to fit into its environment. The standards quoted will be generally those currently recommended by the Ministry of Transport for the design of motorways in rural areas in England,[1] but in some cases the standards for urban motorways[2] will be given also, for the purpose of comparison.

Traffic

Before the design of the motorway can be started, we must know what traffic is expected to use it. This will probably have been estimated as part of the justification described in Chapter 4.

In designing the motorway the engineer must take account of the various stages in the development of the whole road network; allowance being made both for the interim traffic distribution with only the motorway in operation, and for the ultimate traffic distribution when the whole road network is provided. For example, on the Lancashire-Yorkshire Motorway, M.62, between Worsley braided interchange and the interchange at Whitefield, the layout now under construction provides for dual three-lane carriageways, but allows ultimately for dual four-lanes between inter-

[1] Ministry of Transport, *Memorandum on the Design of Motorways in Rural Areas.*

[2] Ministry of Transport, Scottish Development Department, The Welsh Office, *Roads in Urban Areas*, London. Her Majesty's Stationery Office, 1966.

1 Grade-separated roundabout on thc M.6 at A.6, Bamber Bridge, Preston, 29/M6.

2 Broughton interchange, 32/M.6.

3 Barton high-level bridge.

4 Thelwall bridge.

5 Lune bridge.

6 The Gathurst viaduct carries the M.6 over the River Douglas, the Leeds and Liverpool Canal, the Wigan–Southport Railway Line and an occupation road. It is 800 feet long and 87 feet above normal river level, and there are six spans; two end spans of 100 feet each and four 150 feet long intermediate spans. The bridge is 108 feet wide between parapets and, in addition to dual 36 feet wide carriageways, hard shoulders are provided.

7 Lodge Lane bridge carries M.6 over the Liverpool–East Lancashire Trunk Road A.580 at their interchange, into which A.49 also connects. The bridge has eight spans, is 565 feet long, and has a width between parapets of 108 feet to accommodate dual 36 feet wide carriageways and hard shoulders. A feature of the bridge is the boat-shaped reinforced concrete piers cut back to improve visibility for traffic using the roundabout. The universal steel beams in the deck are continuous throughout the length of the bridge and carry a reinforced concrete deck slab.

8 The Samlesbury bridge, the first major bridge to be built on a motorway in Britain, carries the Preston By-pass section of M.6 over the River Ribble, Trunk Road A.59 and an occupation road. A three-span continuous all-welded steel box girder bridge, it has spans of 120, 180 and 120 feet and a width between parapets of 94 feet.

9 The Fylde junction higher bridge at Broughton Interchange is 1,300 feet long and has 11 spans, seven of 130 feet and four of 90 feet 6 inches. The steelwork of the bridge has been painted in two shades of blue. The columns, cantilevers and the underside of the box girder are in dark blue and the vertical faces of the girder are in light blue.

10 The Snowhill bridge, awarded a Class I Commendation by the Civic Trust, is a two-span propped cantilever bridge carrying an Unclassified County Road over the Preston–Lancaster Section of M.6. The road has a very steep gradient averaging 1 in 10 and falling from east to west across the motorway, which accounts for the particular type of design. The reinforced concrete abutments and single pier are founded on rock. The pier is monolithic with the cast-in-situ continuous cellular deck, which is of post-tensioned prestressed concrete. The main span over the motorway is 156 feet and the side span is 81 feet.

11 Parkhead Lane bridge: a single 119 feet span reinforced concrete portal frame bridge carrying a farmer's occupation road and a stream over the Preston–Lancaster Section of M.6. The stream flows in a six feet wide open channel alongside the occupation road. The mass concrete wing walls are founded on rock and faced with split concrete blocks, red in colour.

12 Mount South bridge: one of a pair of typical two-span bridges carrying the roundabout at the junction of the Winwick Link with M.6. The deck consists of universal steel beams carrying a reinforced concrete deck slab in two equal skew spans of 67 feet. The pier has a smooth concrete finish and the abutments are faced with natural stone.

13 Jeps Lane bridge: a three-span cantilever and suspended span bridge with a centre span of 122 feet and side spans of 49 feet. It carries an Unclassified County Road over the Preston–Lancaster Section of M.6. The cantilever and anchor arms are of in-situ reinforced concrete hollow box construction. The 55 feet long suspended span consists of precast pretensioned beams with an in-situ concrete filling.

14 Kenlis Arms bridge: a typical four-span bridge carrying a Class III Road over the Preston–Lancaster Section of M.6. The two main spans are each 64 feet 6 inches long and consist of precast pretensioned beams placed side by side with diaphragms in between. The side spans, each of 38 feet, have the same type of beams but with a solid in-situ concrete infilling. The deck is carried on reinforced concrete bank-seat type abutments founded on the approach embankments and on reinforced concrete skeleton piers.

15 Shaft footbridge: a single-span three-pin arch-type reinforced concrete foot-bridge. The span is 150 feet and the main ribs and decking are of precast units.

16 Lydiate Farm footbridge: a three-span footbridge of the cantilever and suspended span type. The centre span is 112 feet and the two side spans are 50 feet each. The anchor and cantilever arms are precast reinforced concrete and the centre suspended span is of pretensioned precast concrete. The piers are of mass concrete and the abutments reinforced concrete.

17 Example of old mine gallery encountered during excavation.

18 Example of old mine shaft encountered during excavation.

19 Maintenance compound on M.6.

20 Gritting vehicle.

21 Mowing machine developed for use on steep slopes.

22 Charnock Richard service area on M.6.

23 Forton service area on M.6.

changes, when both the Irwell Valley Motorway into Salford and Manchester, and the North-East Lancashire Motorway are operating. When this occurs, the Lancashire-Yorkshire Motorway will be carrying over 80,000 vehicles per day on this length (based on 1979 volumes) and will be one of the most heavily trafficked sections of motorway in England.

For design purposes, the Ministry of Transport recommends that normally the design should be for predicted traffic 20 years ahead at a growth rate of 5 per cent compound interest per annum. The expression 'passenger car unit' (p.c.u.) is used to express the amount of space different types of vehicles take up on the road compared to a car. A bus, for instance, is equivalent to three cars on a rural motorway.

The traffic data obtained should include figures for tidal and peak hour flows, and allowance must also be made for monthly and daily variation of traffic.

Carriageway capacity

The recommended capacities of rural motorways[1] are given in Table 6.

Table 6
Rural motorway capacities

Standard	Average journey length in miles	Daily capacities in p.c.u.'s per 16-hour August day			Peak hourly capacity per lane in p.c.u.'s
		Dual 2 lane	Dual 3 lane	Dual 4 lane	
1	Over 25	33,000	50,000	66,000	1,000
2	10–25	40,000	60,000	80,000	1,200
3	Under 10	50,000	75,000	100,000	1,500

The standards adopted for design depend upon the average journey length. Standard 1 is applicable to the best conditions of operation, but where average journey lengths are shorter, higher capacities can be accepted with reduced speeds. Where details of tidal and peak hour flow are known, the directional peak hour capacities in the table should be used for design.

Standards of alignment

'Sight distance' is the clear distance over which a driver would be able to see an obstruction from an assumed eye level of 3 feet 6 inches above the surface of

[1] Ministry of Transport, Scottish Development Department, The Welsh Office, *Layout of Roads in Rural Areas*, London, H.M.S.O. 1968.

the road. The minimum sight distance along rural motorways is 950 feet in both the vertical and horizontal planes, and should be checked along the centre line of both nearside and offside lanes of each carriageway. This sight distance is equivalent to the minimum stopping distance for a speed of 70 miles per hour if there were an obstruction ahead. This may seem a long distance, but at 70 miles per hour 950 feet are covered in only nine seconds. At interchanges, where visibility is of particular importance, longer sight distances should be provided, if possible.

In the preliminary location of a route, it should be borne in mind that on the inside of a curve of under about 5,800 feet radius, a bridge parapet two feet behind the hard shoulder will reduce the sight distance below the 950 feet minimum. Where a bridge pier is in the central reserve, three feet behind the marginal strips, and the radius is less than 11,000 feet, the sight distance is again reduced below 950 feet. It follows, therefore, that low radius curves should be avoided where possible, but where they cannot be avoided, the effect of the presence of bridges on sight line must be carefully studied. Also it is necessary to take into account any extra cost in widening of the verges or central reserve in order to obtain the required sight distance.

Subject to clearance by the Ministry of Transport, there is some relaxation in the minimum sight distance of 950 feet on the nearside, where the obstruction causes only a momentary reduction of visibility, as at a bridge pier. Here it is still necessary to have the full 950 feet visibility to the rear of the bridge pier and the clear sight distance in front of the pier must be not less than 700 feet, with the length of pier not more than 60 feet, and width not more than three feet. Only one such pier should be within a 950 feet chord, so that the radius of the motorway where there are two bridges for a roundabout over the motorway, needs special study to ensure that this criterion is met.

The effect of the radius of curvature of the motorway on sight distance may be gathered from the fact that for a 950-feet-long line, the mid-ordinate (the distance at the centre between the straight and the curve) is about 11 feet for a curve of 10,000 feet radius and 38 feet for a curve of 3,000 feet radius. If the motorway were in a rock cutting with near-vertical sides, a radius of 3,000 feet would necessitate having the verge width increased from the present five feet standard to 22 feet, that is a further 17 feet. This can result in appreciable increase in cost of the motorway.

In considering the preliminary alignment of the route the motorway designer needs to appreciate the significance of sight distance in relation to the curvature, bridges and economics.

On urban motorways, the minimum sight distance is that appropriate to their design speed. At 50 miles per hour for instance, on a dual carriageway, the minimum sight distance is 425 feet which is the minimum stopping distance at this speed.

Vertical curves are provided at all changes of gradient along the motorway to give a satisfactory standard of appearance, comfort to travellers and visibility. The vertical and horizontal alignment should be co-ordinated with gradual changes, as described later, in order to preserve the flowing alignment of the motorway.

The Ministry of Transport recommends a minimum radius of 60,000 feet for summits and 30,000 feet for valleys and that wherever possible the length of a vertical curve should not be less than 1,000 feet. The minimum sight distance of 950 feet must still be provided if a reduction in radius is contemplated at a summit and this limits the absolute minimum radius at a summit to 30,000 feet.

The normal maximum longitudinal gradient for a motorway in rural areas is three per cent (1 in 33). In hilly country, however, a four per cent (1 in 25) gradient may need to be accepted. It may be possible in hilly country to separate the two carriageways and provide a less steep gradient for climbing vehicles, but this could create a problem if it is necessary to provide access for agricultural or other purposes to the land between the two carriageways.

On long climbing grades an extra lane to the carriageway may be justified, if heavy commercial vehicles form a substantial proportion of the traffic. As a guide, an extra lane may be required on grades of three per cent (1 in 33) over 1,500 feet in length and on a four per cent (1 in 25) grade over 1,000 feet long.

Drainage is the limiting consideration with the minimum longitudinal gradient of the motorway. Modern methods of laying drainage channels and carriageway surfacing enable less steep gradients to be used than formerly and gradients of 1 in 1,000 have been laid without causing any drainage difficulties. However, it is better where possible to work to a minimum longitudinal gradient of about 1 in 250 (0·4 per cent) and this only where no reversal or 'roll-over' in superelevation occurs, as described below.

The horizontal curves on a rural motorway are set out as arcs of a circle and normally the minimum radius is 2,800 feet, but this standard may need to be reduced at certain types of interchanges between two motorways.

To counteract the centrifugal effect tending to turn vehicles over, and to give greater side friction between the wheels and the road surface, it is necessary to superelevate the carriageway. On a straight length of motorway, the slope across the carriageway, or crossfall as it is known, falls outwards from the central reserve and it is the practice for this adverse crossfall to be eliminated on all curves of 11,000 feet radius or less. The rate of superelevation is calculated from the design speed of 70 miles per hour and the radius of curvature; the sharper the radius the greater being the superelevation. The limiting values on superelevation are that it should never be steeper than 1 in 14½ nor flatter than the normal carriageway crossfall of 1 in 40.

For curves of 11,000 feet radius and below, the horizontal circular arcs are provided at each end with a transition curve and the superelevation is applied gradually over this length. This transition curve is a spiral, gradually decreasing in radius over a designed length, from the straight line to the radius of the circular arc. The reason for this is that no vehicle can effect an instantaneous change in direction. It takes time for the driver to turn over the steering wheel and, in order that the alignment may be inherently safe, the entering and leaving of a circular arc must be accomplished gradually at a rate appropriate to the speed of the vehicle, the radius of the curve and the superelevation of the road. The centrifugal force on the vehicle in negotiating the curve is thus applied and released gradually without discomfort to passengers, helping to keep the vehicle in its traffic lane so that it almost steers itself. If the vehicle is travelling at the design speed around a bend, this is sometimes referred to as the 'hands-off' speed–although it is not advisable to try this on a motorway!

The provision of transition curves and superelevation helps to preserve the flowing alignment of a motorway, as well as being essential for the safety of traffic.

When superelevation is applied, it is usual to keep the edges of the central reserve at the normal level and pivot about these edges. The superelevation is applied gradually, so that the nearside edge of carriageway does not vary in longitudinal grade steeper than 1 in 200, relative to the edge of the central reserve. Usually the vertical alignment of these roll-over lengths is designed individually by plotting them to an exaggerated scale with smooth curves, again to enhance the flowing appearance.

This will often mean that the normal crossfall of 1 in 40 outwards from the central reserve edge of carriageway will need to be eliminated on a right-hand curve on the straight before reaching the start of the transition curve where it is usual to aim to have the carriageway flat. For example, the length of transition curve for a radius of 5,720 feet at 70 miles per hour is 250 feet and the superelevation required on the circular arc is 1 in 40. It is necessary therefore, to 'roll-over' from 1 in 40 outwards to 1 in 40 inwards and for a 36-feet-wide carriageway, this represents a total rise for the outer edge of carriageway of 1·8 feet which, if applied at the steepest rate of 1 in 200, would take a length of 360 feet to develop. This is longer than the 250 feet design length of transition curve, so that some of this difference is taken up on the straight. The important point in design is that the full superelevation must be attained by the time a driver reaches the circular arc portion of the curve.

On the lengths where superelevation leads to a situation where both edges of carriageway are at the same level, steps have to be taken to ensure that the motorway is on such a gradient that there is still a longitudinal fall to drain water away from the flat area. If the superelevation is applied at a valley in the motorway profile, rain water could cover the flat area and take a long time to drain away, giving rise to dangerous 'aquaplaning' of the wheels of fast-moving vehicles in wet conditions.

Cross-section standards

The current Ministry of Transport cross-section is given in Figure 21 and shows the standard total formation width to be 92 feet for a dual two-lane motorway and 116 feet for a dual three-lane motorway. The standards previously applicable between 1962 and 1966 were 105 feet and 129 feet respectively, the amendment having been made on economy grounds.

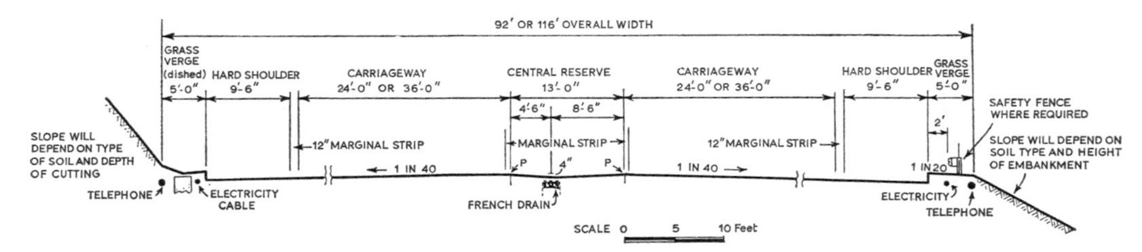

Fig. 21. Current Ministry of Transport motorway cross-section for rural areas

The total width needed for the motorway will include the width for slopes of cuttings or embankments, as well as the formation width itself.

Marginal strips are provided at each edge of the carriageway of contrasting texture and colour. The nearside edge of carriageway is defined further by red reflecting road studs at 54 feet spacing, except at acceleration or deceleration lanes where green reflecting studs are provided at 27 feet spacing along the nearside edge of the main carriageway and red studs, also at 27 feet spacing, along the nearside edge of the acceleration or deceleration lane. The offside edge of carriageway is defined by amber reflecting studs at 54 feet spacing.

The central reserve is a nominal 13 feet wide, grassed over. The hard shoulders on the nearside of each carriageway are paved, and the surface sealed. The practice in Lancashire is to use a bituminous sealing material called 'Schlamme' with an added red pigment to provide additional contrast with the carriageway. The verges behind the hard shoulders accommodate drains, electricity and emergency telephone services, safety fences and signs, so that the width allowed has been put to good use!

The angle of the slopes of embankment and cuttings depends upon the soil encountered as revealed by the soil survey. It must be accurately known at an early stage so that land requirements can be plotted on the land plans and the areas calculated. In cuttings in poor ground it may be necessary to be as flat as 1 in 3 or in some cases even flatter, whereas in solid rock the slopes may be nearly vertical.

The standard carriageway crossfall on the straight of 1 in 40 is continued on the marginal strips and paved hard shoulders as with the carriageway, both on normal crossfall and superelevation.

On the central reserve and verges, where French drains are provided, the crossfall slopes towards this drain. A French drain consists of an open-jointed or porous pipe laid in a trench which is filled with loose stone, forming a filter through which water can percolate.

Blending the motorway into the landscape

The detailed design of the alignment of a motorway requires the artistic application of the prescribed standards of layout, and vertical and horizontal alignment, into the environment and topography through which the route will pass. This inter-relationship between the motorway and its environment is of great importance, for the geometric design of a motorway cannot be divorced from the aesthetic aspect of design.

Due consideration will be given in the design to residential, agricultural and industrial interests and future development proposals, with a view to causing the minimum of interference; a broad, objective view must be maintained so as to cause as little detriment to the surroundings as possible.

The motorway must be pleasing to the eye both of the motorist using it and of the person who views it from its surroundings. It must make provision for adequate inter-connection across the motorway for the communities living alongside. The motorway can open up hitherto unknown vistas of beauty of the area through which it passes and these must be appreciated and allowed for in the design. For example, the existing all-purpose road may pass through the legacy of terraced dwellings of the Industrial Revolution which effectively blot out any views of green fields, whereas the motorway route skirting this area could open to view an area of outstanding natural beauty. Of course, this must be designed with extreme care to ensure that the motorway itself does not spoil the countryside.

Although the designer has a bird's-eye view in plan of the motorway, this is not how the motorist will see it. He needs to bring his eye-level nearer to the paper and view along the route, putting himself in the place of the motorist. For this purpose, models are useful in visualising the overall effect, particularly at complicated multi-level interchanges. Various types of road model equipment are available and are helpful in designing this marrying-in of the vertical and horizontal planes.

In general, it is better for the motorway to have gentle curvature, presenting an ever-changing backcloth to the driver and avoiding the dangerous hypnotic and monotonous effect of extremely long straights. Horizontal and vertical curves should be as long as possible with no short curves or straights. This is particularly important where two curves are both visible to the motorist–they should not be connected by short straights, but should flow one into the other. If two curves follow each other in the same direction, they should be connected by a flat curve rather than a straight. Small changes of alignment should be avoided, as they will give the effect of a pronounced kink.

A straight alignment can be dramatic on occasions, if it draws attention to an outstanding landmark such as a church or tower at the end of the straight. Straights should preferably not be longer than say two miles. A switch-back profile which can be seen from any point along the motorway should be avoided if at all possible by smoothing out the longitudinal profile, or by providing horizontal curvature.

One objection to long straights is that they increase the effect of what is known as 'parallelism'. This results from the impression given by several of the constructional features of the motorway running in parallel, such as lane markings and studs, the edges of carriageway, back of hard shoulder, top of embankment, toe of cutting and motorway fencing. The aim here should be to break up the long-distance parallel effect which is not in accord with the surroundings. Varying the width of central reserve or the level of the carriageways can help to achieve this, but the need to keep land acquisition to the minimum often rules this out. Clump planting of shrubs on the central reserve helps to break up the long view of several parallel lines. Bridges along a straight can also help in this respect.

Where possible the horizontal and vertical curves should both begin at the same point along the motorway. Similarly they should both end at a common point. By this means both horizontal and vertical curvature go hand-in-hand to produce a flowing alignment for the motorway which needs to be considered from the beginning as a three-dimensional problem. This phasing applies especially to horizontal curves of less than about 5,000 feet radius and vertical curves of less than 50,000 feet radius.

The design should aim to provide curves of consistent and similar radii on adjacent lengths of motorway, without an abrupt change of standards. Where there is no option but to provide the minimum allowable radius, because of mountainous country or density of development, it is preferable for the respective curves on either side to be reduced in radius gradually as the minimum radius is reached. This principle is useful at the approach to interchanges also. When a motorist leaves the motorway designed for 70 miles per hour and travels along a slip road which can only be designed for, say 30 miles per hour because of site difficulties, it is helpful if a curve between the two can be introduced which is designed to an intermediate standard, say, of 50 miles per hour. Traffic is then gradually and safely guided into reduction of speed and standards. The American practice of providing a sign at slip roads on interchanges giving the speed value to which the curve has been designed has much to commend it, by informing the driver of the safe speed. The Ministry of Transport is in fact conducting trials on all-purpose roads to investigate the value of signs of this type.

Where the route crosses the skyline, it should preferably be on a curve so as to present a uniform background. The top and bottom of slopes of cutting and embankments should be rounded off where possible and not have a sharp angular ridge. Sharp horizontal curvature should be avoided near a summit because a driver has little advance indication of the change in direction, nor at the valley of a sagging curve as this produces a broken-back effect.

It has been mentioned previously that superelevation of the carriageways should be applied gradually, but in addition each edge of carriageway needs to be designed to 'roll-over' gracefully.

In some isolated circumstances, on sidelong ground for example, or on open moorland, it may be feasible to separate the carriageways or vary the level of one carriageway relative to the other–not just for the sake of effect but only where economically justifiable.

Where there is a sharp switch from cutting to embankment, it is preferable to warp over or flare gradually at the ends of the cutting to avoid the 'cheese-slice' effect. Trees at the end of a cutting can help to frame the view as the motorist emerges on to an embankment.

The landscaping of motorways is sometimes looked upon as the planting of trees after the motorway is constructed, but this is only one limited aspect. Landscaping in its true sense must be built into the design of the motorway from the very beginning and is inherent in the layout. No amount of extensive tree planting can disguise a basically poor layout. Tree planting should provide punctuation marks on the journey, helping to merge the motorway into its surroundings so that it becomes an integral part of the scenery. Sometimes it is possible to dispense with a fence line to a woodland altogether and thus help to bring the motorway and its environment into still greater harmony.

Survey of the route

For the initial selection of a motorway route in Britain, the Ordnance Survey maps of various scales as large as 1/1250th can be used.

When the Ministry has agreed the recommended route, it is necessary to have more details of existing features along the route and the side roads affected. For this purpose, contour plans to a scale of either 1/1250th, or preferably 1/500th scale, are prepared from ground or aerial surveys covering a strip say 600 feet in width along the motorway route. Aerial survey involves not only photographing the route from the air and subsequent plotting to scale of the ground features, but necessitates a great deal of work on the ground as well as physical checking.

The provision of contour plans, giving the existing levels of the ground, is more useful to the engineer designing the route than cross-sections at intervals along the proposed centre line of the motorway. Should there be an alteration in the horizontal alignment of the motorway, the cross-section would need to be levelled again, whereas the contours provide the information for levels needed for any alignment within the width covered. The contour intervals of height can be as close as one foot, which is adequate for the determination of the limits of the land required.

Reference marker points are needed at intervals along the route, from which the motorway can be set out later, and it is useful to provide the National or Local Grid references of these marker points, so that this location can be correlated with the Ordnance Survey references. Also, if all the plans are marked with the intersection points of a grid at say 500 feet intervals, it enables the motorway centre line to be calculated by co-ordinates and plotted more easily to a high degree of accuracy on the various sheets of the aerial survey plans.

The survey of the route calls for accurate measurement of angles, distances and levels. For the measurement of angles, modern theodolites can read to an accuracy of one second, that is 1/3600th of a degree. For the measurement of distance, either tellurometers or geodimeters may be used. Tellurometers measure distance electronically in much the same way as radar. Geodimeters perform the same function by measuring the time taken for a concentrated, modulated beam of light from the instrument to be bounced back by a reflector held at the other end of the distance to be measured.

Soil survey and its influence on design

During the initial investigation in selecting a route for a motorway, general information regarding soil conditions can often be obtained without the need for a soil survey. For example, the memoirs and maps of the Geological Survey of Great Britain give details of solid and drift deposits; the local office of the Soil Survey of Great Britain has details of soil types and depths of peat; the National Coal Board has records of strata and workings in mining areas, and the Mineral Valuer and local

Planning Authority are able to provide geological information. The best way to obtain an overall picture of soil conditions along a route which is under consideration is to walk over every yard of the route, carefully noting conditions which give a clue to the strata beneath. Of course, in some instances, a preliminary soil survey may be needed on sections of possible routes, where conditions are so important as to play a major part in the choice of route alignment.

When all the factors relating to possible routes have been assessed and one selected and approved in principle, the next step is to carry out a detailed soil survey along the route. Its purpose is to provide factual information of soil and groundwater conditions as a basis for rational and economic design, also to predict and cater for any constructional difficulties which may arise. The information required from a soil survey includes:

1. Guidance in the selection of the horizontal and vertical alignment.
2. The materials suitable for compacting in embankments.
3. The safe slopes of cuttings and embankments.
4. Extent and character of rock excavation.
5. Sub-soil and surface drainage.
6. Thickness of the carriageway pavement.
7. The treatment or removal of any weak deposits beneath the pavement.
8. Local materials which may be stablised and used as a construction material in the base of the pavement.
9. The supporting value of strata beneath bridge foundations and the type of foundation to be provided.
10. The presence of any harmful sulphates in the groundwater which may attack concrete structures.
11. The consolidation characteristics of the ground underlying the embankments so that precautions can be taken to reduce settlement.

The soil survey involves borings and trial pits, taking samples on site, and the examination and testing of these samples, supplemented in some cases by seismic or resistivity surveys. The most commonly used boring equipment is the mobile mechanical percussion boring rig, with normal diameters of from four to six inches. Lining tubes are used as the bore deepens, to retain the sides of the bore and to keep groundwater from seeping down the bore and affecting the soil samples as they are taken. Various chisels and cutting heads are used to break up the soil at the bottom of the hole where necessary and at intervals of about five feet in depth, or closer, the disturbed soil is removed and undisturbed samples are then taken in four inch diameter tubes, 18 inches long. For rock, rotary rock-drilling equipment is usually needed, unless the rock is very soft. The cutting head may need diamond tips if the rock is hard.

Foundations can be affected by mine workings, landslips, shrinkage or long-term settlement of clay soil, soil movement due to frost or moisture change, lack of lateral support, over-stressing of strata, faults, and earthquakes. The site exploration should provide such information where relevant, so that the necessary precautions can be taken.

Although some appreciation of the differences in behaviour of various soils under loading have been known for centuries, it is only comparatively recently that foundation problems have been approached scientificially, with an understanding of the basic properties of the soil.

The coarser grained soils, gravel and sand, are roughly cubical or spherical in shape, and form a honeycomb pattern, like oranges in a box, with direct contact between the particles. These contribute to soil stability by virtue of their property of internal friction at the points where they touch. Gravelly and sandy soils are open, well-drained non-susceptible to frost action, non-plastic and when a load is applied any settlement is almost immediate.

The next size down is the silt, which has similar physical and chemical properties to sand, but its particles are too small to be seen with the unaided eye. The main contribution of silt to soil stability is the internal friction, but with a certain amount of cohesion or stickiness also. Silt has particular susceptibility to frost damage through the formation of ice lenses during periods of prolonged frost. These ice lenses accumulate in horizontal layers and are formed because the permeability and particular size of the voids in silt are favourable to the capillary attraction of water into the upper layers in much the same way as blotting paper soaks up ink. The formation of these ice lenses can result in heaving of the surface.

Clay minerals in soil have the smallest particle sizes and are plate or rod-shaped with a flocculent structure, rather like soap flakes in a carton. This structure gives the key to the intractability of clay and is a main cause of its property of plasticity or cohesion. Clay contributes to soil stability through this property of cohesion. It has no internal friction when wet, the particles being separated from each other by a lubricating film of water of microscopic thickness which, at the same time, binds the particles together by surface tension and other forces. Clay soils are poorly drained and because of this are not susceptible to frost action, but this slow drainage means that settlement upon application of a load takes place over a long period of time. This is termed consolidation and gives rise to particular difficulties with bridges, where damage to the abutments and decking can occur, particularly if the effect of settlement is not uniform over the area of the foundations. The stability of clay slopes is dependent upon the height of slope as well as the angle of slope and their design requires to be checked by 'slip-circle' analysis to ensure that the factor of safety is adequate. The term 'slip-circle' is used because a mass of clay has the characteristic of failing by slipping along a plane shaped as an arc of a circle.

For general foundation design, soils may be divided into five main groups: (i) rocks; (ii) non-cohesive soils (i.e. granular or sandy in nature); (iii) cohesive soils (i.e. sticky or clayey in nature); (iv) peat (i.e. organic in nature); (v) made or filled ground. As a guide to the range of the supporting value of various soils, the maximum safe bearing capacity for a spread foundation on rock ranges from 100 tons per square foot for hard rock such as granite in sound condition to six tons per square foot for solid chalk. Non-cohesive and cohesive soils range from six tons to half a ton per square foot or less. In addition, the cohesive soils need to be checked for long-term consolidation settlement. With peat or other highly compressible soils, the safest method is to remove it altogether, but this is not always economically possible. Many soils, of course, are a mixture of gravels and sands, silts and clays, and an infinite number of combinations of these materials occur in their natural state.

Types of pavement

A road pavement is the term given to the various layers of material placed on the surface of the ground after the completion of earthworks. Its main functions are, firstly to distribute the traffic loads over the surface of the ground and protect it from the adverse effects of the weather, and secondly to provide a smooth running surface.

There are two main types of pavement construction–flexible and rigid. In flexible construction the running surface consists of a layer of bituminous material. Underneath is the main structural element known as the base. Below this, additional support is provided by a layer of granular material referred to as the sub-base, placed directly on the surface of the ground or sub-grade as it is known.

In rigid construction a concrete slab is the main component. Not only does it provide the basic structure of the pavement but the running surface as well. It is invariably necessary to lay a base of granular material on which the slab is cast. The essential difference between a rigid and flexible pavement is that a rigid pavement possesses a strength in tension, whereas a flexible pavement is assumed for design purposes to have no tensile strength.

A third type of pavement involves the use of concrete as the base material in what would otherwise be a flexible form of construction. The concrete used in this type is of comparatively low strength, and as it contains only a small amount of cement, it is known as 'lean concrete'. It provides a pavement which is neither fully flexible nor wholly rigid. It is known as semi-flexible or composite construction.

Flexible and semi-flexible construction

Most methods of pavement design are either empirical or partially so. One of the best known is the California Bearing Ratio or C.B.R. method, which has been used in Britain for some time. The test which is carried out, in order to arrive at the C.B.R. value, was developed by the California State Highways Department many years ago, hence the name. It consists of determining the load/penetration relation for a plunger of three square inches cross-sectional area which is forced into a sample of soil prepared in a standard manner. The Department found that a certain crushed stone required a load of 3,000 lb. to push in the plunger by 0·1 inch and this was taken as the standard C.B.R. value of 100 per cent. A soil which needs 300 lb. to push in the plunger by 0·1 inch thus has a C.B.R. value of 10 per cent.

It was found that a material of a certain C.B.R. value required a certain minimum thickness of construction for a specified wheel load. As a result of tests carried out on materials of known performance, sufficient data was obtained to enable design curves to be plotted. From these it is possible to extract the depth of construction necessary for a particular C.B.R. value.

The C.B.R. values for soils in Britain may vary from 0 for poorly drained, very weak clays and silts, up to 60 per cent for well drained sandy gravel. It is usually necessary to remove material with very low C.B.R. values, and if this operation is carried out to a shallow depth, it in effect lowers the formation and increases the thickness of sub-base.

Where large pockets of poor quality material have to be removed, it is clearly uneconomic to back-fill with sub-base material. Either selected excavated material or imported granular material would therefore be used, care being taken to ensure proper compaction.

The volume of traffic which a road is required to carry is a major factor affecting its design. In the case of the pavement, the heavier vehicles will obviously have the greatest effect. An estimate is made of the number of commercial vehicles of unladen weight exceeding 30 cwts., which will travel over the road in both directions 20 years after the road has been constructed.

The Ministry of Transport has prepared its own Design Charts, using the C.B.R. method, and these were contained in the first issue of Road Note 29, *A Guide to the Structural Design of Flexible and Rigid Pavements for New Roads*, published in 1960.

In the subsequent issue of 1965[1] there are six Charts relating to different traffic categories. Design Chart 1, which covers the highest category of more than 4,500 commercial vehicles per day, is used for motorways.

[1] Ministry of Transport, *A Guide to the Structural Design of Flexible and Rigid Pavements for New Roads*. Road Note 29, Second Edition, London, H.M.S.O. 1965.

The basic requirements of this Chart are that a four-inch thickness of two-course bituminous surfacing shall be laid on a base of ten-inch maximum thickness. The thickness of sub-base varies in relation to the C.B.R. value of the sub-grade. For a value of two per cent a thickness of 20 inches is required, but for five and a half per cent only six inches is considered necessary. This means that the total depth of construction within this range of values will vary from 34 inches (20 + 10 + 4) down to 20 inches (6 + 10 + 4).

If exceptionally good sub-grade conditions are encountered with C.B.R. values of 30 per cent and over, it is theoretically possible to dispense with the provision of a sub-base completely. There are, however, other factors to consider. As the depth of frost penetration in Britain may be as high as 18 inches–even more in severe winters–it is essential for any material within 18 inches of the surface to be non-frost susceptible. If this condition is not complied with, then frost heave may occur with the possible break-up of the carriageway under traffic.

In addition, a sub-base is a satisfactory means of providing a drainage layer in the lower level of the pavement. It also provides a working platform on which constructional equipment is able to operate and which can be accurately shaped to receive the base. The sub-base itself must have a C.B.R. value of not less than 30 per cent.

Various alternative types of base may be used. The ten-inch thickness may be made up of three inches of dense tar or bitumen macadam, or rolled asphalt, laid on seven inches of lean concrete.

Alternatively, a fully flexible base of reduced depth may be adopted and either eight inches of dense tar or bitumen macadam, or seven inches of rolled asphalt may be used. In these cases it is necessary to maintain the full depth of construction required by increasing the thickness of sub-base by two inches and three inches respectively.

The four-inch thickness of surfacing has a wearing course of one and a half inches laid on a two-and-a-half-inch thick base course. The wearing course is of hot rolled asphalt with precoated chippings rolled into the surface to give a high resistance to skidding. The base course, however, may be of hot rolled asphalt or of dense bitumen macadam or tarmacadam. Although the same basic principles of using the C.B.R. method were employed in the construction of the early motorways in Britain, such as the Preston By-pass, less expensive materials were used in the base.

Rigid construction

Road Note 29 contains Design Charts for concrete pavements based on three types of sub-grade.

'Very susceptible to non-uniform movement' is used to describe organic soils, heavy clays with low C.B.R. values, and sub-grades with pockets of peat within a depth of 15 feet below the surface.

'Very stable' sub-grades are those with well-graded sandy gravel.

'Normal' describes sub-grades between the two extremes defined above.

For motorways in Britain the slab is generally reinforced concrete but the use of unreinforced concrete is under consideration. According to the Design Chart the thickness of slab may be 10, 11 or 12 inches and the base thickness below the slab 0, 3 or 6 inches, depending on the sub-grade.

Base materials may be similar to those used for sub-base in the construction of flexible or semi-flexible pavements. Lean concrete or soil cement, which are described in Chapter 9, may also be used. Where the base is to carry construction traffic, which is generally the case in motorway construction, it is recommended that the base thickness be increased by three inches.

It is important for the condition of the base to be such that it provides a uniform condition on which to lay the slab. This will not be the case if it is damaged by construction traffic and it is necessary to treat the depths quoted as minimum requirements.

The minimum weight of reinforcement is given in the Chart as 9 lbs. per square yard. It is required to conform with B.S.1221 'Steel Fabric for Concrete Reinforcement' and to have 2½ inches of cover from the surface of the slab.

Longitudinal joints are to be provided so that the slabs are not more than 15 feet wide. It is usual to provide such joints at the spacing of the carriageway lanes, i.e. 12 feet. To prevent the longitudinal joints opening, tie bars must be provided across these joints or the arrangement and weight of the mats of reinforcement must be designed to perform this function.

The maximum spacing of expansion joints is 240 feet, reducing to 160 feet, when the pavement is laid in cold weather. Contraction joint spacing is restricted to a maximum of 80 feet.

Concrete unfortunately shrinks, particularly during its early life, and tensile stresses are developed. Although strong in compression, it is weak in tension and, unless the stresses are controlled, cracking will occur. The purpose of the contraction joints is to relieve the stresses by allowing the slab between two adjacent joints to contract and, therefore, they must be positioned at fairly close intervals. The reinforcement is incorporated in the slabs between the joints to resist any remaining stresses and, as far as possible, prevent cracking. Should cracks occur, however, the reinforcement will prevent them opening and allowing water to enter. There is, therefore, a direct relationship between the spacing of the contraction joints and the amount of reinforcement.

The joints may be made by either forming a groove in the wet concrete or by making a saw cut as early as possible in the hardening concrete to produce a plane of weakness. A crack will then develop below the groove or saw cut, but it is necessary to provide a form of load transference across the joint from one slab to another at the mid depth of the slabs and dowel bars are incorporated for this purpose.

Expansion joints are incorporated to relieve compressive stresses and require the provision of a compressible band of material through the full depth of the slab. Load transference must also be provided.

Both types of joint are sealed on the surface, but it is extremely difficult to find an entirely satisfactory material capable of accommodating the movement involved, particuarly for the expansion joints.

Expansion joints are wider than contraction joints, more difficult to form, more susceptible to the ingress of stones and other incompressible materials, and liable to breaking away at the edges. They are largely responsible for the past criticism of the riding quality of concrete roads. The Ministry of Transport, however, is now considering whether approval can be given to the omission of expansion joints when concreting is carried out between May and October.

In 1962, before starting the construction of the Preston-Lancaster section of M.6, I visited the U.S.A. to examine the methods of constructing concrete roads. I concluded that the practice in the majority of the States, whereby expansion joints are omitted, could reasonably be adopted in Britain. Although it was not possible to reach agreement on this particular point, the spacing was increased and many other design features were adopted on this length of motorway, as described in Chapter 10.

Drainage

The importance of efficient, properly maintained drainage in road construction cannot be over-emphasised. In 1823 John Loudon Macadam put forward his ideas on the basic principles of good roadbuilding, which were that it is the native soil beneath the road which really supports the weight of traffic, and that while it is kept in a dry state with a waterproof covering over it to preserve it from penetration of water, it will continue to carry this weight without sinking. This fundamental principle applies equally to modern motorway construction, as it did over a hundred years ago.

The drainage of a motorway has several functions: to take away rainwater falling on the road; to keep the road construction and soil beneath free from water; to intercept water from the land falling towards the motorway; and to intercept water from the motorway slopes so that it will not cause flooding of adjacent land. The drainage design must also allow the existing watercourses and drainage systems which are crossed by the motorway to operate freely, without either adversely affecting the existing drainage of the area or endangering the stability of the motorway.

Firstly, then, a system of drainage must be provided to take the surface water from the large impermeable areas of the carriageways and hard shoulders, in order to prevent dangerous ponding of water on the surface in heavy rain. This must be conveyed to the existing watercourses which cross the motorway and discharged into them without causing flooding downstream. Close liaison is required with the relevant River Authorities, Drainage Boards and the Land Drainage Department of the Ministry of Agriculture, Fisheries and Food, from the very beginning when a route for a motorway is under consideration, as this could influence the selection of the route. It is preferable that the pipe outfalls conveying motorway drainage should discharge into main watercourses, as these will be better able to cope with the flow. The regrading, realignment, or widening of these watercourses is sometimes necessary downstream of the motorway to take the additional discharge from it.

At the point of discharge, careful design is necessary to prevent scouring of the watercourse and on large outfalls, the flow from the motorway should be brought in smoothly in the direction of flow of the existing watercourse, by means of training walls on both sides and a paved bottom between them.

Where possible, these pipe outfalls should discharge into the main watercourse above its highest flood level. When this is not feasible, allowance must be made in the drainage design of a pipe outfall for the fact that at times of heavy rain the end of the outfall will be submerged, reducing the flow from the outfall.

On earlier motorways, the drainage from the impermeable road surface was collected by porous drains located at the edge of the hard surface. This caused the risk of weakening of the carriageway foundation near the edge through seepage of water from the porous drain. The current practice is to provide for surface water to be collected in channels at the edge of the paved surface, then into gullies and conveyed through a piped drainage system. The size of the pipes can be calculated from recommended design methods prepared by the Road Research Laboratory.[1] It is usual to design the size of the pipes on the basis of a 'one-year' storm, that is, a storm of an intensity and duration which on average would occur once a year, the pipes being designed to take this flow without the backing-up of water into the manholes. The surface water pipes are located deep enough to be safe from damage by both construction traffic and traffic using the motorway, and of such a depth as to allow the porous French drains to be connected into them. A check is made during design

[1] Road Research Laboratory, *A Guide for Engineers to the Design of Storm Water Systems*, Department of Scientific and Industrial Research, Road Research Note 35, London, Her Majesty's Stationery Office, 1963.

to ensure that water will not back-up into the French drain system when a heavy storm occurs. A 'ten-year' storm may be taken for this purpose.

French drains are provided where the motorway is in a cutting, to intercept soil water and surface water down the slopes of the cuttings, also to lower the level of the groundwater, known as the water-table, under the motorway. They are also provided in the dished central reserve. At the bottom of embankment slopes, and at the top of cutting slopes where the land falls towards the motorway, French drains or intercepting ditches are provided. French drains in the verges and central reserve should be at least two feet below the level of the underside of the carriageway foundation.

In the construction of the smooth ribbon of a motorway through the irregular humps and hollows of the topography provided by nature, the natural movement of the water in the soil is disturbed and it takes some time before it settles down again and attains its new equilibrium. Thus the earthworks tend to encounter sources of groundwater and these are more pronounced in the initial stages. Even so, groundwater emerging at points in cutting slopes can persist, and in such cases stone drains will be needed to carry this water into the French drain in the verge of the motorway.

The effect of the motorway works on existing land drainage systems needs to be considered. If any field drainage systems are severed by the motorway a separate cut-off drain should be provided as accommodation works, on the farmer's side of the motorway boundary. This enables him to maintain or extend his drainage system at a later date if he so wishes, on his own land.

The bridges and culverts designed to carry the motorway over existing watercourses should be sufficiently large to carry flood water through them without causing flooding upstream and to safeguard the motorway works. This requires careful investigation of the catchment area of all the watercourses upstream of the motorway, the expected intensity of rainfall and, from this information, the calculation of the expected flows. The design of the clear waterway area of these structures needs to be able to cope with even catastrophic storms where there may be a risk of the motorway works being washed away if there is any backing-up of flood water. Usually culverts across the motorway are at least three feet six inches in diameter to assist in maintenance and it is preferable to collect a number of ditches together and take the flow across in one large culvert, rather than having a separate culvert for each ditch.

Chapter 7

Interchanges

Introduction

In Chapter 1, the main characteristics quoted for a motorway included the complete separation of opposing traffic and the elimination of crossings at the same level. In order to comply with these requirements all motorway interchanges cross at different levels and are termed 'grade separated', involving the provision of one or more bridge structures to carry crossing traffic over the other streams of traffic.

A fundamental principle of motorway design, applicable equally to other roads, is that the traffic capacity of the motorway can be no greater than its capacity at intersections. This principle applies not only to the actual design of individual interchanges, but also to their frequency and location which must be such that the free flow of through traffic is not impeded.

The choice of the type of interchange depends upon many factors, which include the character and volume of traffic on each of the intersecting routes and on the connecting links between them, the design speed for the various traffic movements, topography, land availability and overall cost. The final layout for an interchange is usually evolved after consideration of several designs and often it is necessary to strike a compromise between a layout which results in an acceptable reduction of speed for each element of turning traffic, and which does not take up too great an area of land. The economic aspect also is important and the capital invested in the interchange must be appropriate to the traffic using it.

The location of the various types of interchanges in Lancashire which are illustrated in this chapter can be found in Figure 31 on page 174.

The guiding principles of design

The main routes through an interchange should be so designed that there is no appreciable variation of speed standard from that immediately before and after the interchange. Turning movements must be accommodated safely. For turning traffic at an interchange there will be inevitably some degree of lowering of design standards below those of the motorway proper. Whether it be in design speed, alignment, visibility or width, the lower standard must be appropriate to the volume of traffic using each particular link of the interchange and must be coupled with adequate warning to the driver. Current practice in the U.S.A. is for the connecting links for turning traffic to be designed preferably for a speed of about 0·85 of the design speed of the motorway itself, with a minimum of about 0·4. For a motorway designed to 70 miles per hour, a design speed of 60 miles per hour is preferable for turning traffic and a minimum of 30 miles per hour, with corresponding radii of 1,040 feet and 230 feet respectively.[1]

Ample length must be provided for change of speed on acceleration and deceleration lanes. Where possible the gradient of the connecting roads should be such as to assist in change of speed. This means an uphill gradient on leaving the motorway and a downhill gradient on entering it. The traffic leaving and entering the motorway should do so on the left or nearside of the through traffic. In order to obviate additional weaving, it is preferable that turning traffic should leave the motorway at a point earlier than that which brings traffic on to the motorway at an interchange.

Spacing of interchanges

There is no precise formula for the spacing of interchanges, for each motorway must be treated on its merits. The aim is for the interchanges to provide sufficient accesses to traffic without interfering with the flow of through traffic on the motorway. The Ministry of Transport says that 'in sparsely populated rural areas interchanges spaced 10 to 12 miles apart may suffice for normal traffic and for police, maintenance and emergency requirements, but in more developed areas, closer spacing will be necessary. The minimum spacing in rural areas should not normally be less than 3 miles'.[2]

On the M.6 in Lancashire, the longest length between interchanges is 13½ miles between Broughton and Hampson Green on the Preston-Lancaster section, but on this length there is a private access for police, maintenance and emergency services, which can also gain access through the Forton Service Area.

On the other hand, on the six-mile Stretford-Eccles Motorway, M.62, there are four intermediate interchanges between the two terminal junctions, giving an average spacing of about 1·2 miles between interchanges. This motorway carries over 30,000

[1] American Association of State Highway Officials, *A Policy on Geometric Design of Rural Highways*, Washington, D.C., 1965.

[2] Ministry of Transport *Memorandum on the Design of Motorways in Rural Areas.*

vehicles per day and collects traffic from the adjacent densely developed areas, including the Trafford Park Industrial Estate, which employs more than 50,000 workers. Although this motorway is only dual two-lane layout, little difficulty is at present experienced either by traffic using the interchanges or by that proceeding on the main motorway.

The minimum length for the weaving of traffic between the entry taper of one interchange and the exit taper of the next, governs the absolute minimum spacing for interchanges. This weaving length depends upon the speed and volume of the weaving traffic. Taking as an example traffic weaving at 30 miles per hour on an urban motorway, for a weaving volume of 1,000 p.c.u.'s per hour, a minimum weaving length of 250 feet is required but, where the volume rises to 3,500 p.c.u.'s per hour, the minimum weaving length increases to 1,850 feet.[1]

Types of interchange

The different types of motorway interchange can be divided into certain basic types:

Two-level interchanges:

A. Diamond
B. Split diamond
C. Grade-separated roundabout
D. Partial cloverleaf
E. Trumpet
F. The free-flow terminal
G. Full cloverleaf

Three-level interchanges:

A. The T free-flow terminal
B. Grade-separated roundabout with two through routes

Four-level interchanges:

A. Fully-directional
B. Semi-directional

Two-level interchanges

The various types of two-level interchanges described below are shown in diagrammatic form in Figure 22.

[1] Ministry of Transport, Scottish Development Department, The Welsh Office, *Roads in Urban Areas*, London, Her Majesty's Stationery Office, 1966.

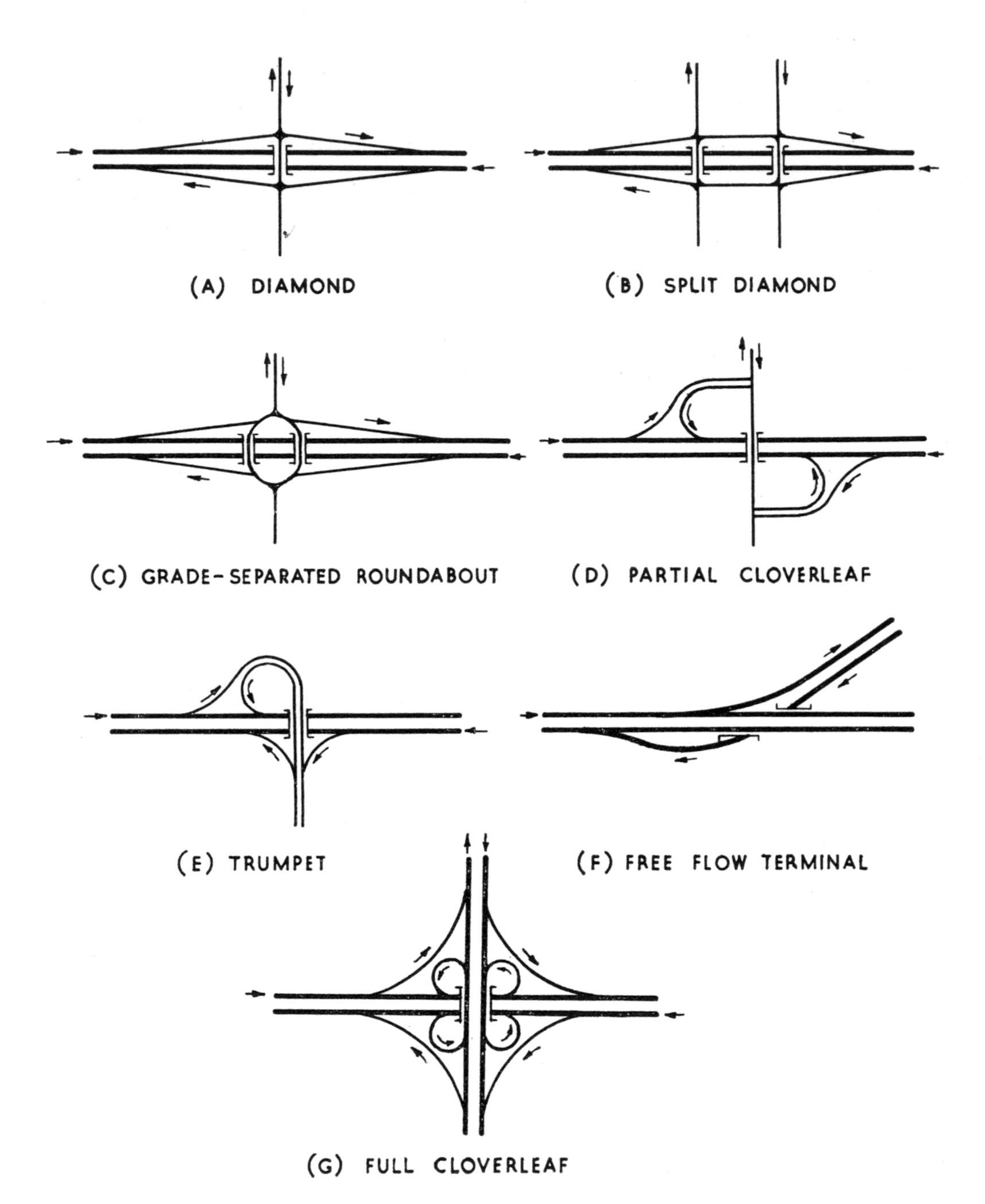

Fig. 22. Two-level interchanges

A. Diamond

The diamond is the most simple type of layout and is used where there will be only a light volume of traffic interchanging between the motorway and the all-purpose road and where the crossing all-purpose road has also only a light volume of traffic. Access for emergency, police and maintenance purposes could be added justification for this type of interchange.

This interchange has the advantage that the initial cost additional to that of the motorway is relatively low compared with other types of grade-separated interchanges. Also the land requirement is low, being about three acres in addition to that needed for the motorway.

The diamond interchange would be appropriate for an all-purpose road with a design volume for 1979 of up to say 10,000 vehicles per day and up to 3,000 vehicles per day entering or leaving the motorway. The interchange also causes little interference with traffic flow on the motorway, traffic for the interchange having left the motorway before the traffic enters from the interchange, with a reasonably long length between the exit and entrance tapers.

The main disadvantage of the diamond type of interchange is the danger in crossing the traffic on the all-purpose road at one level, immediately after leaving a motorway with no crossing traffic. The right-turn movement on leaving the slip road is potentially the most dangerous movement at this type of interchange. Thus a sudden alteration in standards can catch the unwary driver by surprise. Right-turning traffic should be required to halt if there is appreciable traffic on the all-purpose road, but this could result in queueing of vehicles on the slip road and under severe peak traffic flow the queue could stretch back on to the motorway and result in most dangerous conditions.

Careful attention is needed to the layout of the junction between the slip road and the all-purpose road and the provision of guide and refuge islands on the all-purpose road should be considered as should also the provision of traffic lights.

B. Split diamond

The split diamond interchange is a development of the diamond type. It would be applicable under the circumstances described above, but with two roads crossing the motorway. Preferably, these roads should each be one-way.

C. Grade-separated roundabout

An example of this type is the interchange No. 29/M.6 between the M.6 and A.6 at Bamber Bridge, Preston, shown in Plate 1.

The roundabout, which is situated on the all-purpose road, may be either above or below the motorway route. As mentioned previously, it is desirable to have an uphill gradient on leaving the motorway and, therefore, the roundabout should preferably be above the motorway.

This type of interchange would be provided where a higher volume of turning traffic would use the interchange than with the diamond type and where there is sufficient land available to construct the roundabout. The interchange requires two bridges. The land required additional to that for the motorway would occupy about ten acres. The slip roads, as with the diamond type, have a smooth alignment with no sharp curves; they do not interfere to any appreciable extent with motorway traffic and have a good distance between the exit tapers for traffic leaving the motorway and the entry tapers for traffic coming on to the motorway.

The roundabout itself is designed to accommodate the traffic movements on the same principles as with a surface roundabout. The grade-separated roundabout can be adapted to accommodate additional all-purpose roads at the interchange, provided that all those traffic movements can be accommodated on its weaving sections. Where the all-purpose traffic needs to maintain a relatively high speed, the grade-separated roundabout would not be appropriate, unless a third level for straight-through traffic is added, as described later.

The grade-separated roundabout can also be modified to provide for a T junction with an all-purpose road.

A development of a grade-separated roundabout is given in Figure 16, showing the interchange between the Lancashire-Yorkshire Motorway at Bury New Road, A.56, in Whitefield, where pedestrians are catered for, clear of traffic, by means of four subways and a footbridge, in the centre of the roundabout.

D. **Partial cloverleaf**

The partial cloverleaf, shown in Figure 22, provides in essence a junction with the all-purpose road similar to the diamond type and has the same disadvantage of danger to right-turning traffic on the all-purpose road. This can be overcome by providing roundabouts at the junction of the slip road with the all-purpose road. It has the further disadvantage of needing more land and having relatively sharp curves on the slip roads. Preferably the outer slip roads, with the easier curves, should cater for the traffic leaving the motorway, the slower-moving traffic from the all-purpose road being on the inner loop.

It is current practice for the slip roads, where parallel, to be dual carriageways, although, on some earlier motorways, the slip roads were two-way. As with the diamond and grade-separated roundabout, vehicles leave the motorway in advance of the vehicles entering it, but the length between the exit and entrance is shorter and this is a disadvantage. Under some site circumstances, the partial cloverleaf layout may be the mirror image of that shown in Figure 22.

There is a modified partial cloverleaf interchange at Samlesbury, Preston, on the M.6 motorway, No. 31/M.6, designed to avoid an obstruction on one side of the interchange. At this location, the obstruction was the River Ribble, running parallel to the all-purpose road.

E. **Trumpet**

The trumpet type of interchange, shown in Figure 22, can be used where the all-purpose road, or motorway spur, forms a T or Y junction with the motorway. The layout should be designed so that the heavier turning traffic volumes are carried on the easier curves, with the lighter turning traffic volumes on the inside of the loop. Also the inner loop should preferably be on entry to the motorway, rather than on leaving the motorway, on which vehicles have been able to sustain high speeds for long distances.

The trumpet pattern is widely used as a basis of interchange design and many variations are possible, depending upon site conditions, and which traffic flows are to be favoured. The land needed is about 12 acres for an inner loop designed for 30 miles per hour speed, but the area would rise sharply for increased design speeds.

Where the turning traffic volumes are heavy and require a more free-flowing alignment, as a motorway-to-motorway connection, an interchange involving three levels may be required.

F. **The free-flow terminal** (Movement in two directions)

Where one motorway ends and joins another motorway, a free-flow terminal, as in Figure 22, can be provided. This needs only one bridge structure, provided that traffic movements are required in two directions only. The land required is about 12 acres. If, however, traffic movement is required in all three directions, then a third level at the interchange is introduced.

G. **Full cloverleaf**

The full cloverleaf interchange, also indicated in Figure 22, is applicable to four-leg interchanges, that is, two crossing routes. It keeps the two intersecting main streams of traffic entirely separate and provides for all possible movement of traffic in any direction at the interchange. The potentially dangerous right-turning movement across a stream of traffic is eliminated by taking the turning traffic safely over or under this stream of traffic, then providing a loop road through 270 degrees to merge it on the nearside of the traffic which the turning traffic desires to join. The left-turning traffic is catered for by the outer loops, which can provide a high standard of alignment.

It follows, therefore, that the conventional standard cloverleaf favours left-hand turns, but right-turning traffic around the inner loops is to a much lower standard. It is possible, however, to introduce modifications to the cloverleaf to provide a smoother flowing alignment to one or more of the right-turning movements.

One inherent serious disadvantage of the basic cloverleaf interchange is that, by its very layout, weaving of traffic is involved, because vehicles join the motorway at an entry taper before traffic leaves the motorway at an exit taper. This disadvantage can be overcome by the provision of collector-distributor slip roads parallel to the motorway, on which these weaving manoeuvres can take place clear of the main stream of traffic, but this of course adds to the cost. Another disadvantage is the large land requirement, upwards of 50 acres, depending upon the standard of design of the inner loops.

For a rural motorway with a design speed of 70 miles per hour and working to a desirable design speed of 60 miles per hour for the slip roads, a radius of 1,040 feet would be required. The minimum design speed for the slip roads is 30 miles per hour, requiring a radius of 230 feet. With the four 270-degree loops of a cloverleaf, a very large area of land is required to cater for the desirable design speed, and there is considerable extra travel distance for right-turning traffic. Because of this, the loops are seldom designed for a speed above 35 miles per hour.

Three-level interchanges

A. **The T free-flow terminal** (Movement in three directions)

As mentioned under F above, if a three-way interchange is needed to cater for traffic movements between all three directions, a third level is required. Figure 23 shows the diagrammatic plan of such a layout. The area of land required would be about 40 acres.

B. **Grade-separated roundabout with two through routes**

In this type of interchange, shown in Figure 23, the roundabout is placed in between the two motorways so that one passes above it and the other below, and are connected to it by diamond-shaped slip roads.

This design needs either a viaduct and two bridges, or five bridges, but its capacity for turning traffic is limited to that of its roundabout. An example is shown in Figure 24 at the proposed interchange between the Lancashire-Yorkshire Motorway, M.62, and the North-East Lancashire Motorway, also named the Bury Easterly By-pass.

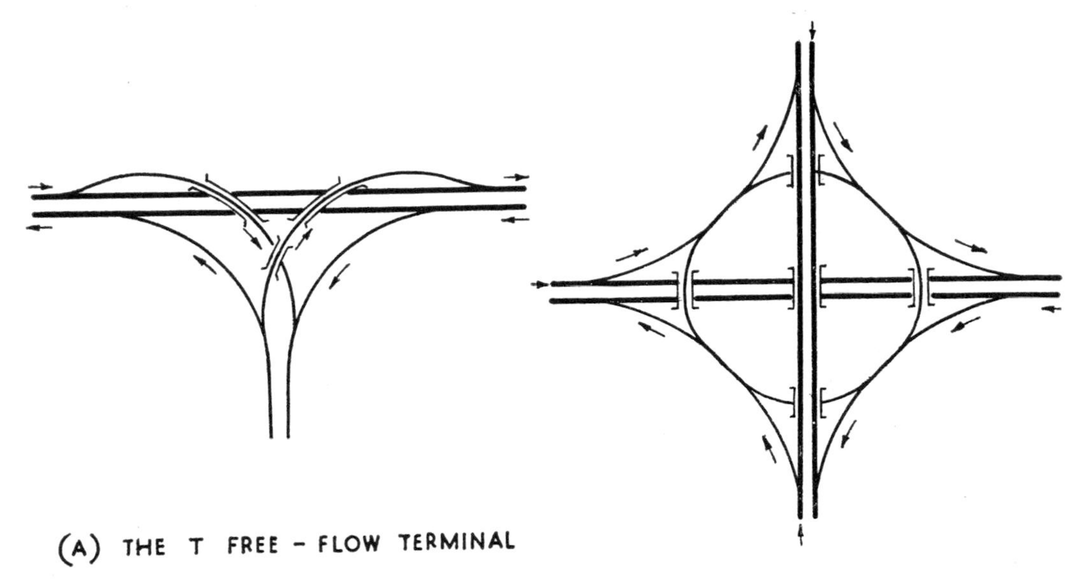

Fig. 23. Three-level interchanges

Fig. 24. Interchange between Lancashire–Yorkshire Motorway, M. 62, and North-East Lancashire Motorway

The land required for this type of interchange without allowing for a free-flowing movement is about 15 acres.

Four-level interchanges

A. Fully-directional

This type of interchange is shown in Figure 25 and provides a high standard of layout for heavy volumes of through and turning traffic on major motorways. A

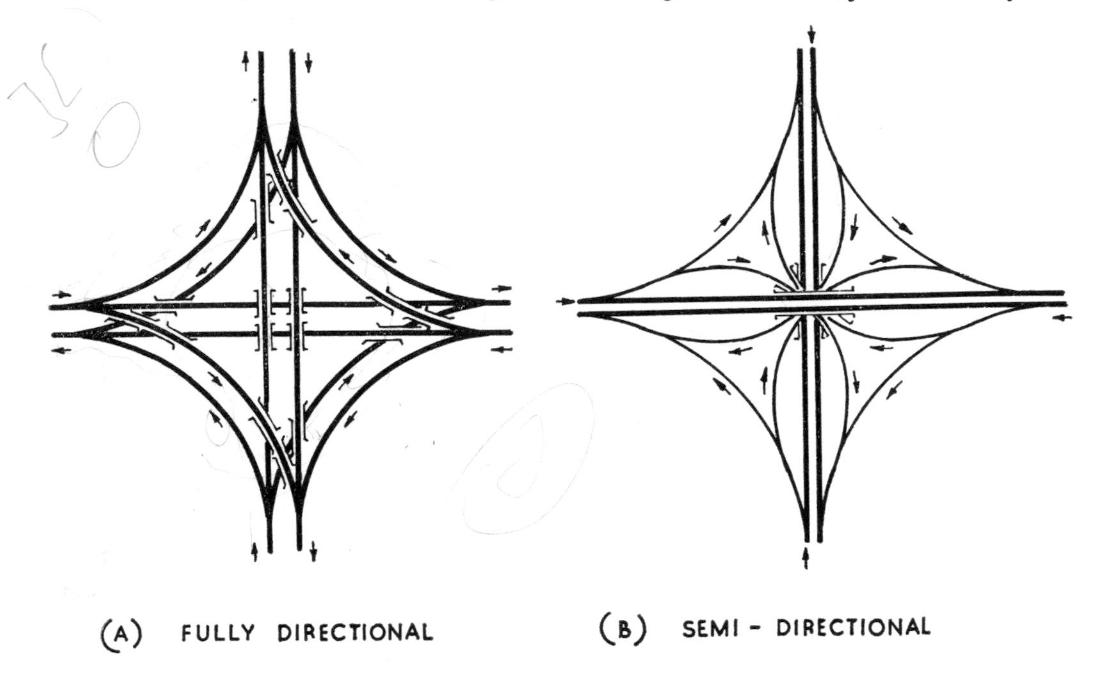

Fig. 25. Four-level interchanges

directional interchange is one which is designed to provide free-flow direct connections for major right-turning traffic movements. If the interchange is fully directional it provides these connections on the side and in the direction which the traffic wishes to go. In other words, right-turning traffic leaves on the right of the main traffic stream and left-turning traffic on the left. Leaving on the right is not normal policy and can be disconcerting for the driver.

There are four different levels to the interchange. These are the two levels of the intersecting routes with one of the direct connections above and another direct connection beneath, making four levels in all. The interchange needs a total of 16 bridges.

I am not aware of any interchange of this type being constructed in Britain.

B. Semi-directional

The semi-directional four-level interchange, also shown in Figure 25, overcomes the disadvantage of traffic leaving on the right. The term semi-directional is applied to interchanges where the free-flow direct connections for major right-turning traffic movements always take place on the left, or nearside, of the main stream of through traffic.

An example of this type is the Almondsbury interchange, north of Bristol, between the M.4 and M.5 motorways. An earlier example is the interchange at Los Angeles, California, which is said to be the most heavily trafficked interchange in the world, carrying over 300,000 vehicles per day.[1]

The four-level design involves reverse curves on the connecting roads for right-turning movements. The four-level structure at the centre of the interchange will impose a tremendous pressure on the strata beneath the foundations, so that rock, or extremely high strength soils, are desirable at the site, otherwise foundation costs could be very high. The area of land occupied is about 60 acres.

The procedure of design

The many factors involved in arriving at a final layout for an interchange cannot always be evaluated precisely. However, there are certain steps which need to be followed in design, which have been put neatly in an official American publication,[2] and I quote from it below:

'1. Obtainment and analysis of traffic data to determine design hour volumes for all through and turning movements, including future expansion.
2. Obtainment of physical data for the site, including maps showing topography and culture, and plans showing existing buildings and those likely in the future.
3. Determination of the location, type and general design features of all highways and other development, both existing and planned, in the area which may have a bearing on the design.
4. Preparation of study sketches for several likely intersection schemes that are suitable to meet traffic needs and are practical for the site and design controls.
5. Analysis of alternate schemes and selection of the better two or more for further study and for preparation of preliminary plans and profiles.

[1] John Hugh Jones, *The Geometric Design of Modern Highways*, London, E. & F. N. Spon Ltd., 1961.

[2] American Association of State Highway Officials, *A Policy on Geometric Design of Rural Highways*, Washington, D.C., 1965.

6. Preparation of preliminary plans and profiles for the alternates selected under 5.
7. Evaluation of each alternate preliminary plan with respect to design features, capacity versus volume, operational characteristics, overall adaptability, maintenance of traffic during construction, and suitability to stage construction.
8. Calculation of preliminary cost estimates for each alternative preliminary plan, including land acquisition, clearing the site, construction, maintenance, utility changes, maintenance of traffic during construction, etc.
9. Calculation of road user costs and road user benefit ratios for each alternative preliminary plan.
10. Joint analysis of values from steps 7, 8 and 9 to reach conclusions as to the preferred plan.
11. Final design, including preparation of construction plans, specifications and estimates.'

Some examples

To illustrate the design procedure, how five examples have evolved is described below.

A. **Broughton interchange** (No. 32/M.6)

The 13½ miles of motorway between the Preston and Lancaster By-passes is the fourth and final length of the M.6 motorway in Lancashire on which the County Council acted as agents for the Minister of Transport for the design and the supervision of the works.

The route extends from a point on Preston By-pass at Broughton, approximately one mile from the northern end of the By-pass, to a point on Lancaster By-pass, about a quarter of a mile from its southern end at Hampson Green, and connections to the existing road system via the trunk road A.6 are provided only at Broughton and Hampson Green.

The interchange at Broughton is the first three-level interchange to be completed on a British motorway. The problem here was to design a layout for the interchange which would cater initially for traffic connecting to the Trunk Road A.6 north of Preston, and ultimately for the large volume of traffic between the Fylde Coast and M.6 when the link between the M.6 and A.6 is extended westwards as a motorway.

The Broughton interchange, therefore, was required to be a motorway-to-motorway interchange, capable of taking safely high traffic volumes, for which the more conventional two-level layout would have been inadequate and have prevented continuous flow.

The design had to allow for: (a) marrying in to the existing Preston By-Pass with the minimum inconvenience to traffic already using the motorway; (b) ensuring that the levels were such that the lowest carriageway could drain into the nearest main watercourse, Blundell Brook, which crosses the interchange site; (c) keeping the area of land taken to the minimum; (d) providing for the culverting or diversion of the watercourses in this area; (e) keeping the load of embankments to the minimum because of the very poor ground conditions in the area; and (f) ensuring that the whole interchange was pleasing to the eye and landscaped into its surroundings.

The reasons for the layout adopted as the most suitable solution are summarised below, and the layout is shown in Plate 2.

Traffic movements at this interchange would be such that at peak periods the flow on the motorway link to the coast would be greater than on the north-south M.6 route. It was necessary, therefore, to provide for the traffic from the north to Blackpool to travel on its own link clear of other traffic streams and enter on the nearside of the much greater volume of traffic which will be travelling from the south to Blackpool, thus providing satisfactory merging of traffic. This traffic movement is catered for by means of Fylde junction higher bridge which passes over the top of four other traffic streams. Traffic from Blackpool to the south passes in a smooth curve under the north-south route, which is carried over by Fylde junction lower bridge.

An additional traffic problem which had to be overcome was to provide traffic entering and leaving the interchange with satisfactory lengths for acceleration and deceleration. This was attained by reducing the dual three-lanes on the M.6 to dual two-lane carriageways through the interchange. By this means traffic both from Blackpool to the north and from Blackpool to the south was provided with a clear traffic lane on entering M.6. Overhead gantry signs erected on M.6 ahead of the junction direct traffic for Blackpool into the nearside lane, with through traffic on M.6 into the middle and offside lanes.

The minimum design speed for the turning traffic links is 45 miles per hour with corresponding visibility distances, which led to the adoption of inclined pier legs to Fylde junction lower bridge, in order to provide adequate sight distance around the curve.

The bridge design was considered at the outset as an integral part of the basic layout of this interchange.

It was arranged that Fylde junction higher bridge would be on a continuous curve of 1,146 feet radius throughout its length and that the whole viaduct would have an aesthetically satisfying elevation when viewed from all approaches. (See Plate 9.)

Twin-column steel box-section piers carry the main steel spine girder, of box-section with steel cantilever arms, supporting the reinforced concrete deck. This construction was adopted for economy, appearance and rapidity of design, the time available being extremely short, with less than a year between the decision on the layout and going out to tender.

The design of Fylde junction lower bridge called for some ingenuity, the inclined legs allowing adequate sight distance, as referred to above, and at the same time helping to keep the construction depth to the minimum, as this in turn affected the level of the Fylde junction higher bridge.

The alluvial material in the flat valley of Blundell Brook, which crosses the site of the interchange, was deep and soft. It was not economical nor practicable to remove this and, in order to keep subsequent settlement to the minimum, the motorway embankments were formed in pulverised fuel ash which is much lighter than ordinary filling.

Even so, some settlement was inevitable and to overcome this problem Blundell Brook culvert was constructed of twin eight feet diameter concrete pipes in ten feet lengths, with flexible joints (the largest size of concrete pipe made up to that time). Also, the contract stipulated that the embankments should be formed in the first fifteen months of the thirty-two months' contract period, with the carriageway construction of flexible materials to be deferred until the last three months of the contract.

The whole interchange was considered from the landscaping aspect at an early stage. A contour plan of the interchange was prepared, showing the earthmoving required. The aim was to provide flowing and graceful contours between the various routes, with gradual warping of land slopes so that the final layout had a natural shape without any sharp, geometrical or artifical appearance to the ground.

The various routes at the interchange were set out on the site by means of light scaffolding tubes and crossheads at an early stage of design, so that the overall effect could be studied.

B. Worsley braided interchange

This is an interchange virtually between three motorways and two all-purpose roads, one of which is a trunk road of great importance. The motorways are the Lancashire-Yorkshire Motorway, M.62, the Manchester-Preston Motorway, M.61, and the Bolton to the South Motorway. The trunk road connected to the interchange is the Liverpool-East Lancashire Trunk Road, A.580, and the other all-purpose road is the Manchester-Bolton Road, A.666.

The volumes of traffic and complexity of movements presented a formidable geometric design problem (Figures 26 and 27). The route here had been protected from development for several years so that there was little interference with property.

However, the route passed through a large peat moss area, crossed two separate mineral railway sidings and extensive engineering problems also arose.

The advantages of the final layout are summarised below:

The interchange as designed will carry a total of approximately 160,000 vehicles per day in 1979 and provide free-flow for the many complex traffic movements with

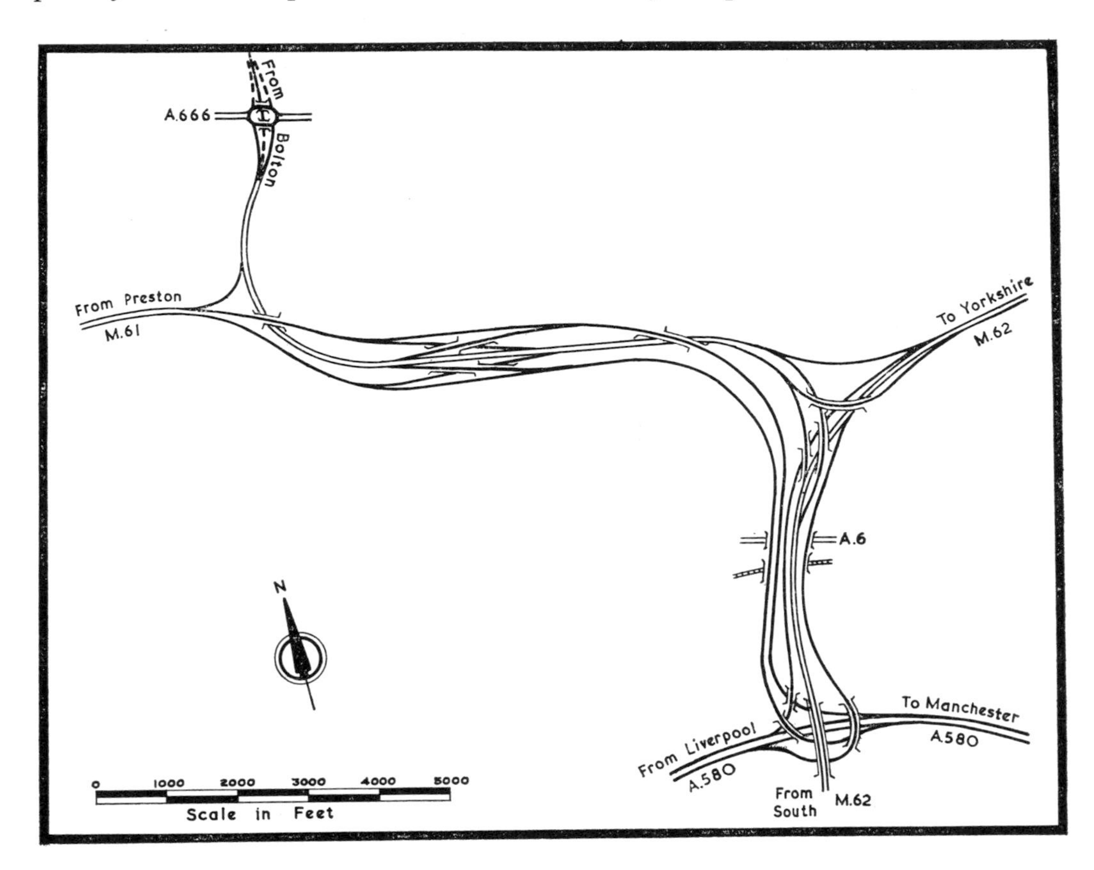

Fig. 26. Worsley braided interchange between M.62, M.61, A.580 and A.666

safety. Direct connections are provided between the Lancashire-Yorkshire Motorway, M.62, and the Manchester-Preston Motorway, M.61, in all directions, requiring this part of the interchange to be three-level. Grade-separated direct connections are provided for the links carrying the heavy volume of traffic between M.61 (Preston) and A.580 (Manchester), between A.580 (Liverpool) and M.62 (Yorkshire) and between M.61 (Manchester) and the Bolton to the South Motorway.

The links between M.61 (Preston) and A.580 (Manchester) carrying these high traffic volumes need not join M.62 so they are kept physically separated from M.62.

To avoid heavy weaving, they are also entirely separated from the A.580 (Liverpool) –M.62 (Yorkshire) links. In order to avoid weaving on the M.62 itself, the A.580 (Liverpool)–M.62 (Yorkshire) links are connected to M.62 north of the M.62 (Stretford)–M.61 (Preston) links.

The relatively light flow of traffic from M.61 (Preston) to A.666 at Kearsley is catered for by a simple left-hand connecting link, but the much higher cost of bridges and connecting links in the reverse direction is not justified. This traffic will be able to gain access to the M.61 for Preston at the next interchange towards Preston, about 1½ miles away.

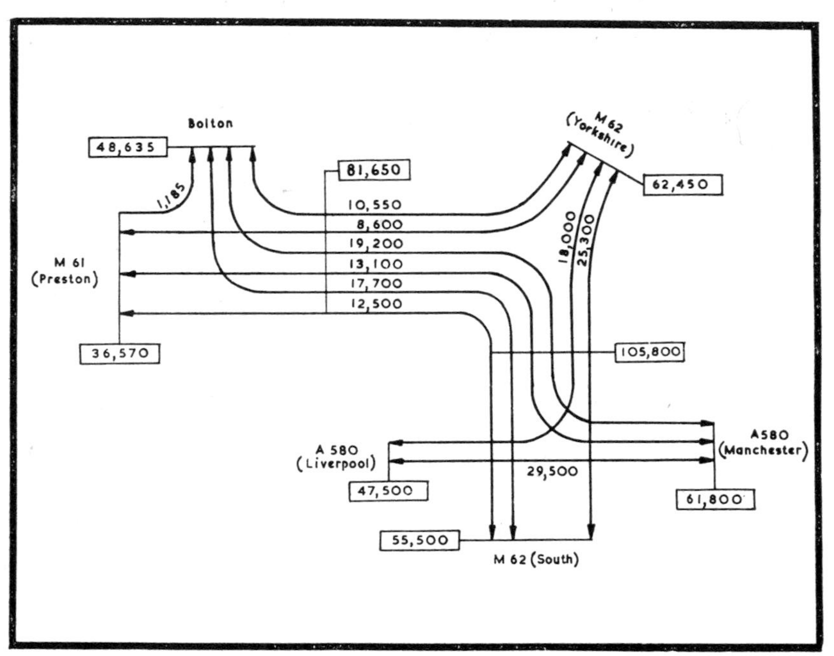

Fig. 27. Worsley braided interchange.
Traffic figures-vehicles per day in 1979

At the interchange the general principle of lighter flows of traffic on the left is followed.

These are the advantages of the layout, but if the final design should appear over-elaborate, a simple comparison with a typical motorway interchange is worthwhile. A section at such an interchange across a dual three-lane motorway and two slip roads would give a total of ten lanes. A section across the Worsley complex, between the M.61/M.62 Interchange and A.580 would give only an additional four lanes, that is, one pair of slip roads. When this is seen in relation to the estimated volume of over 100,000 vehicles per day in 1979, which will use this part of the complex, it will be appreciated that the facilities and capacity provided are no more than necessary.

Although the plan gives a bird's-eye view of the whole interchange, the motorist will be travelling at ground level with large signs to guide him into the route he wishes to take. He will be entirely separate from the traffic on the other routes of the interchange and faced with only one decision at a time, with ample time and distance before coming to another divergence of routes.

Detailed examination was made of several different profiles along the various routes in order to keep the cost down to the minimum. The earlier designs on the north-easterly side of the interchange allowed for passing over a mineral railway. This would have entailed the excavation of 20 feet or so of peat and its replacement with suitable earth fill, followed by further filling to form the embankments. However, it was found that the bottom of the peat could be drained by the construction of an outfall about half a mile in length and the motorway profile was redesigned to pass under, instead of over, the railway. In other words, part of the interchange was lowered to below peat level and, by this means, over two million cubic yards of imported filling saved.

The landscaping aspect has been considered from the beginning, again providing gradual shaping of the ground between the various routes.

C. Interchange between South Lancashire Motorway and Stretford-Eccles By-pass, M.62

The existing Stretford-Eccles By-pass at the present time is being extended northwards and eastwards as the Lancashire-Yorkshire Motorway. The proposed South Lancashire Motorway, between Liverpool and Manchester, will cross the Stretford-Eccles By-pass at a point about one mile south of its present northerly termination.

Referring to the Liverpool-Manchester and Lancashire-Yorkshire Motorways at a British Road Federation Conference on the theme 'People and Cities', held at Liverpool in January 1965, I expressed the opinion that these two routes could revitalise the north of England by providing motorway facilities in an east-west direction from Liverpool and the sea on the west through to the M.1 and Leeds and eventually Hull to the east, thus linking a belt of towns of high industrial importance.

This interchange, therefore, is required to cater for the intersection of two heavily trafficked motorways, one of which is already in operation. It will also carry a high proportion of traffic on two of the link roads, for the traffic between Liverpool and Yorkshire.

The location of the interchange was governed to a great extent by the relative positions of existing interchanges on the M.62 and the possible routes for the South Lancashire Motorway.

Several designs were carefully considered before arriving at the final layout, as shown in Figure 28.

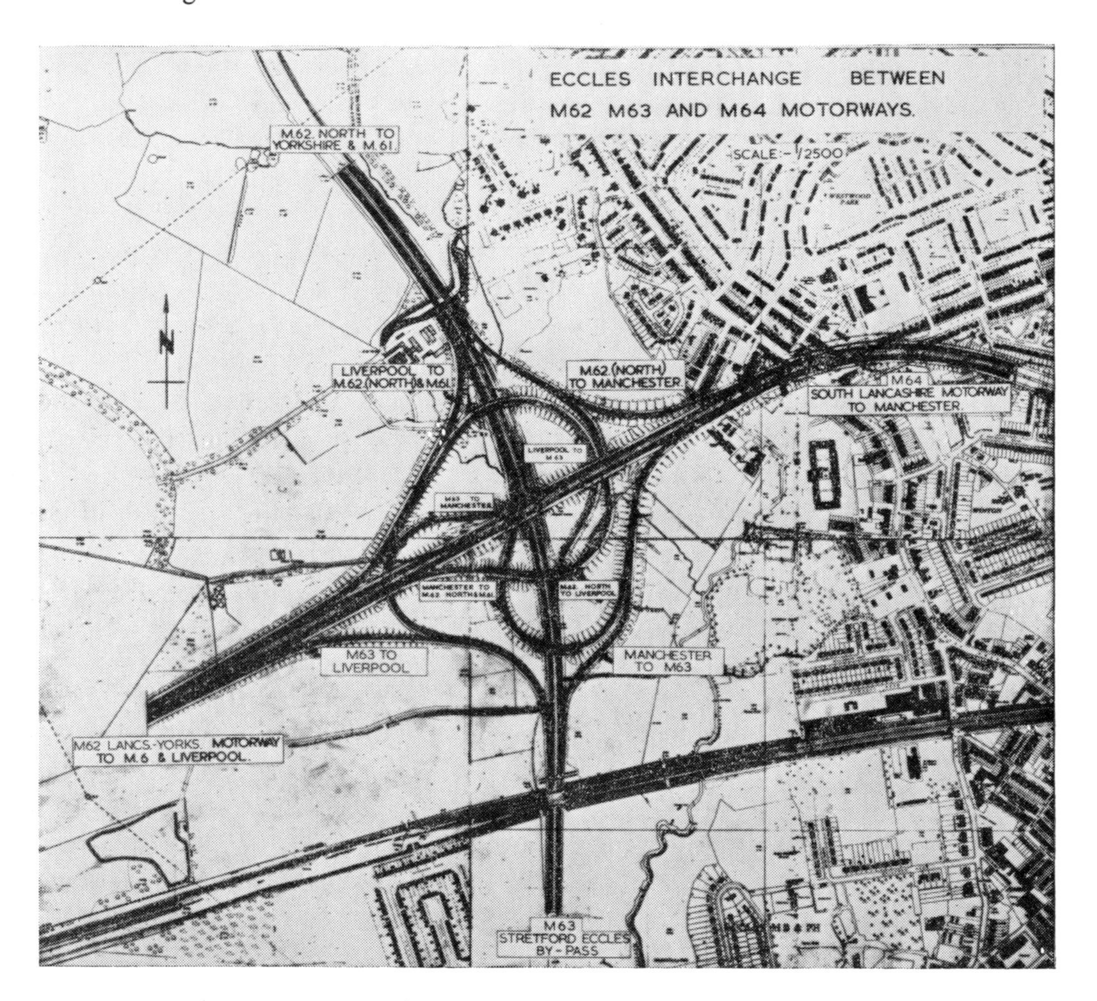

Fig. 28. Eccles interchange between South Lancashire Motorway and Stretford–Eccles By-pass, M.62

D. M.62/M.6 Interchange

This interchange will be sited at Croft, north-east of Warrington, where the proposed Lancashire-Yorkshire Motorway, M.62, will cross the existing M.6 Motorway. The M.6 Motorway at this location runs roughly at ground level and is at the lowest possible level to enable it to be drained to a watercourse. The M.62 Motorway, therefore, and the connecting links at the interchange, have to be above the level of the M.6 Motorway and the braided layout shown in Figure 29 has evolved. By means of this braiding, the amount of filling is kept to the minimum and each intersection of the various routes is at two levels only.

One design which was considered and rejected on the grounds of economy was the four-level semi-directional interchange.

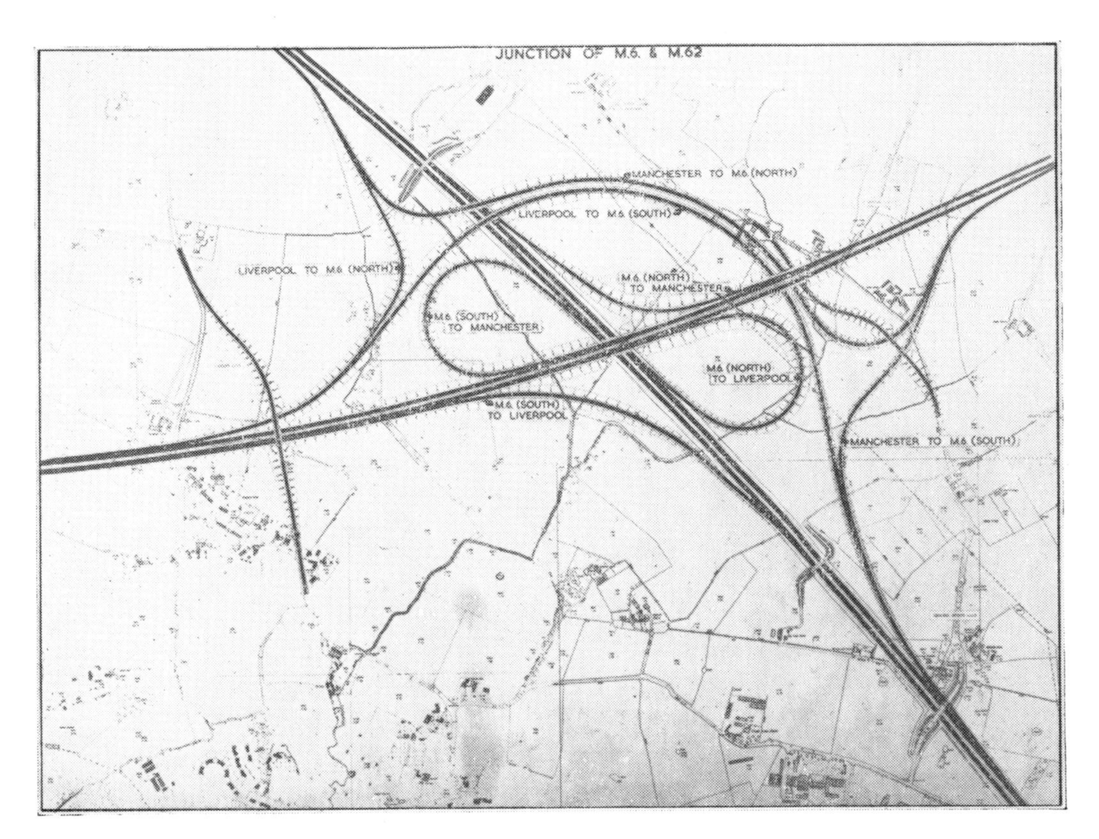

Fig. 29. Croft interchange between South Lancashire Motorway and M.6

A major turning traffic movement at this interchange will be Liverpool to M.6 southbound and in the reverse direction and the layout provides easy curves for these movements.

Although it covers a greater area than the Eccles interchange already described in C above, access to two larger enclosed areas east of the M.6. Motorway will be made possible, enabling the land to continue in agricultural use.

E. M.62/Liverpool Outer Ring Road Interchange

This interchange at Tarbock, east of Huyton, is basically a three-level grade-separated roundabout with two through routes. The interchange has a total of seven legs and provides for a total of 42 traffic movements, those between the Lancashire-Yorkshire Motorway, Liverpool Outer Ring Road, the principal all-purpose Cronton Road, A.5080, and the classified Windy Arbor Lane.

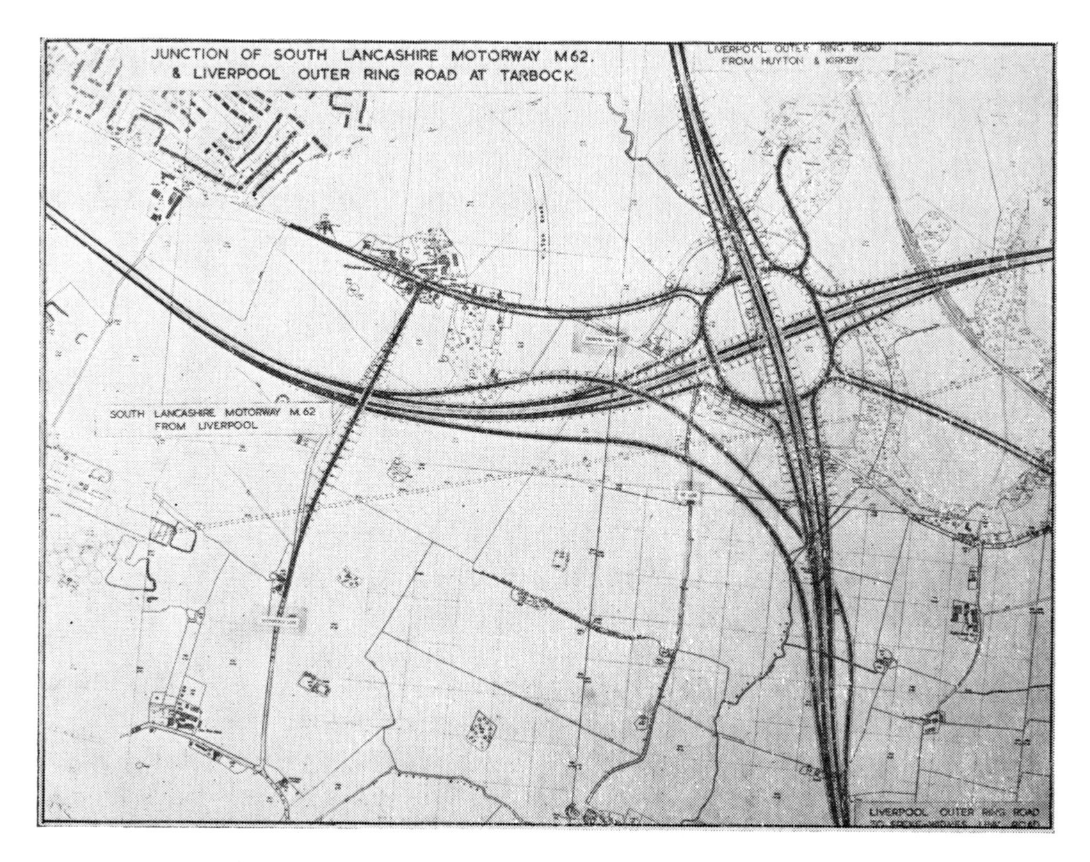

Fig. 30. Tarbock interchange between South Lancashire Motorway and Liverpool Outer Ring Road

The layout is shown in Figure 30. Of the three levels of the interchange, the Lancashire-Yorkshire Motorway is at the lowest level, the roundabout carrying the all-purpose road and turning traffic is at the middle level and the Liverpool Outer Ring Road is carried over the top of the roundabout. A major traffic movement will be from Liverpool to Liverpool Outer Ring Road southbound and in the reverse direction. Free-flowing links will be provided for these movements at a later date, as shown in Figure 30.

Chapter 8

Bridges

The principal features

Ever since man began to travel along primitive tracks he has been faced with building bridges across natural obstacles. The early road builder had limited engineering knowledge and little in the way of tools and equipment; consequently he invariably chose the route of his road so that the bridges would be of the simplest type. In most countries, therefore, the siting of bridges has been one of the most important factors in establishing the existing road system.

In the past, highway construction has largely been confined to the improvement and minor realignment of these old roads and the majority of bridge work has been the replacement of old bridges to accommodate increased traffic volumes and heavier loads.

The building of motorways has necessitated the provision of new bridges on a vast scale to meet many and varied requirements. The basic requirement of complete control of access means that all existing rights of way, whether roads or footpaths, have to be carried over or under the motorway. Also, it is inevitable that the alignment will cut across farm land with some severance of farm units and occupation bridges or cattle creeps must be provided to satisfy the affected farmers.

The design of interchanges connecting with other motorways or existing roads requires the provision of many bridges. In urban areas the motorway might be elevated on viaduct for an appreciable length over buildings, roads, railways and rivers.

The bridges can be divided into two main types: underbridges which carry the motorway and overbridges which pass above it. The distinction is important in relation to the type of loading carried by the bridge. The basis of design loading in Britain is the Standard or HA Loading–defined in British Standard 153–which simulates conditions experienced on a road carrying traffic at full capacity. This loading approximately represents the effect of three 22-ton vehicles closely spaced, followed by ten-ton and five-ton vehicles. For short spans the effect of two nine-ton wheels opposite each other and three feet apart are considered.

In addition, a heavy type of design loading, known as Type HB, represents a single abnormally heavy vehicle, with four axles in two pairs, and four wheels on each axle. The load on each axle is defined as a number of units, each representing one ton and 45 units are applied to give the design loading for a heavy load route. Thus the four axles carry 180 tons.

Motorways are designated as heavy load routes which means that apart from being capable of carrying the Standard of HA Loading the design of all underbridges is checked against the 45 units of HB Loading.

Overbridges carrying public roads are designed for the Standard Loading checked against the appropriate number of HB units depending on the class of road, unless the road is another motorway or heavy load route. Occupation bridges are only designed for half the Standard Loading and footbridges for a load of 100 lbs. per square foot.

A bridge must not only be capable of carrying the superimposed or live load but also its own weight which may form a substantial portion of the total loading.

The length of span is an important consideration in bridge design. The longer the span the greater will be the depth of construction in order to carry the load. It follows that for a given length of bridge, a series of short spans should be more economic than a single long span, as far as the deck construction is concerned. The saving in the cost of the deck will be offset by the provision of the intervening piers which may be costly if ground conditions require expensive foundations.

It is usually desirable for the depth of construction to be kept to a minimum for reasons other than the actual cost of the bridge itself. The required clearance of the bridge over the obstacle to be crossed is always predetermined. For example, a minimum headroom of 16 feet 6 inches over public highways is required in Britain. Similarly, standards are also specified for the crossing of railways. For navigable rivers and canals, satisfactory clearances to accommodate shipping are necessary. Bridges over other rivers or watercourses must provide an adequate waterway to ensure an uninterrupted flow which may present special problems where a motorway crosses the flood plain of a river.

With a specified clearance, any saving in the overall height of the bridge can be made only by reducing the depth of the superstructure. The surface level of the deck determines the height to which filling on the approaches to the bridge will have to be raised. A saving of as little as one foot in the overall height can reduce the requirement of filling by a considerable amount, particularly where a wide valley has to be crossed.

For an overbridge the number and length of spans employed may be as follows:

1. A single span with abutments sited immediately behind the side verges (see Plate 11).
2. Two spans with abutments as above and a centre pier sited in the central reserve (see Plate 12).
3. Three spans in which piers are sited behind the side verges. The centre span crosses the full width of the motorway and the side spans accommodate the slopes of the cutting or the approach embankments (see Plate 13).
4. Four spans, as (3) above, but with a centre pier in the central reserve dividing the main span (see Plate 14).

Whereas in all cases the length of the centre spans will be standard, the side spans in (3) and (4) will vary depending on the width required for the slopes.

With underbridges, the arrangement of spans will vary depending on the particular circumstances. The only instances where any degree of standardisation of lengths of span might apply is in the crossing of railways or of roads laid out to standard widths.

Traffic considerations may affect the arrangement in that the provision of piers in certain positions may seriously affect sight distances, particularly in the case of bridges at interchanges where reduced radii of curvature may be employed.

The choice will depend on two important factors:

1. The type of structural design employed which may or may not be influenced by the materials of construction selected for use.
2. Aesthetic considerations.

The two are closely related but the first essential is that the bridge should be suitable for its purpose from the engineering point of view.

Materials most commonly used in bridge construction are steel and concrete, or a combination of both. Steel has the advantage of being capable of resisting both tensile and compressive stresses. Concrete while good in compression has little or no value in tension. By incorporating steel within the concrete to take up the tensile stresses, reinforced concrete is produced capable of utilisation in any structural member.

A technique employed in more recent years is pre-stressed concrete. In this case the steel is employed to pre-compress the concrete prior to being submitted to loading. Any tensile forces imposed on the concrete have, therefore, to overcome the pre-compression before tensile stresses are developed in the concrete. The amount of pre-compression applied should be sufficient to accommodate the maximum tension developed under the worst loading conditions.

In reinforced concrete the steel, consisting usually of round bars, is merely placed in position in the mould or forms and the concrete cast around it. In pre-stressed concrete two systems are employed–pre-tensioning and post-tensioning. In pre-tensioning, the steel in the form of high tensile steel wires is tensioned prior to placing the concrete. When the concrete has reached a sufficient strength the load on the wires is released. The bond which has then developed between the steel and the concrete transmits the stress from the steel to the concrete and pre-compresses the latter.

In post-tensioning, ducts are cast into the concrete through which steel cables or bars are later threaded. By applying a tensile load to the steel and anchoring it at the ends of the concrete member a pre-compression is applied. The principle can be used in forming very large units by post-tensioning small units together with cables or bars passing through them in line.

Types of bridge

There are several basic types of bridge design. The most elementary is the simply supported span, where the superstructure of the span rests freely on the abutments or piers. A bridge may consist of any number of simply supported spans, each acting independently. This type of bridge is easy to build, but there is an economic limit to the length of span beyond which the depth of construction becomes excessive and the appearance is seriously affected.

For short spans the form of construction can be a simple reinforced concrete slab or it may consist of steel girders or reinforced or pre-stressed concrete beams carrying a reinforced concrete deck slab (see Plate 12). The steel girders may be either standard rolled steel joists or, for the longer spans, specially fabricated plate girders made up from plates welded or riveted together.

Present practice is to connect the beams or girders to the deck slab so that they act together, thereby reducing the cost. Shear connectors are used for this purpose which may consist of studs or steel sections welded on to the top of the steel girders, or bars projecting out of the concrete beams. The deck slab is then cast around the connectors and acts compositely with the girders or beams in carrying the loads.

If a bridge has more than one span and the superstructure is made continuous instead of separated over the supports, the effect of loading is reduced by distribution throughout the spans. The depth of the superstructure can thus be reduced with a saving in cost (see Plate 7). It is important, however, that with this type of design there should be no unequal settlement of the supports. Where this is likely, as in areas subject to mining subsidence, provision can be made for jacking on the supports to maintain the superstructure at the correct level.

For continuous construction, materials similar to those used in the simply supported form of construction can be employed, with the exception of pre-tensioned pre-stressed beams. Steel girders need to be spliced together after erection in order to provide the longer lengths required. Reinforced concrete and post-tensioned pre-stressed concrete require a considerable amount of site work in the form of temporary support whether a simply supported or continuous type of design is used.

For longer spans a cantilever and suspended span type of construction is often adopted and is particularly suitable for three-span bridges over a motorway (see Plate 13). The side spans, known as anchor arms, cantilever into the main span over the piers. A simply supported span is carried on the tips of the cantilevers and forms the centre section of the main span. A greater depth of construction is required over the piers. A pleasing effect can be achieved by providing a curved profile to the underside of the superstructure in order to give a variable depth of section. Steel girders can be used with this type of design. Alternatively, the anchor arms and cantilevers can be formed in reinforced or post-tensioned pre-stressed concrete, in which case the suspended span may conveniently be of pre-tensioned pre-stressed beams.

As they perform the same function, similar types of abutments and piers may be used for any of these designs. The abutment is required to support the superstructure and the live load to be carried by the bridge and also to retain the filling in the embankment approach to the bridge. The latter effect can be reduced by providing additional spans and sloping down the filling within the end spans. In this case the abutment may be of an open type consisting of columns supporting a beam on which the end of the superstructure rests. The filling passes between the columns and the horizontal load applied to the abutment is thereby reduced.

The simplest abutment is the mass concrete wall, but the reinforced concrete abutment has several advantages: it is more economical in materials and lighter in weight, which may be important with inferior ground conditions. It requires, however, more skilled labour and supervision during construction.

Not having to serve as retaining walls, but merely to carry the loads from the superstructure, piers are much simpler to design than abutments. Although piers carry largely vertical loads, certain horizontal loads arising from wind, traction and expansion movement must be provided for. Piers can be comparatively slender and may be either in mass or reinforced concrete. Occasionally steel columns of circular or rectangular section are used.

In an ordinary simply supported span the maximum effect of loading occurs at the centre of the span. If, however, the ends are firmly held the effect is reduced and a saving in depth of construction can be achieved by making the superstructure monolithic with the abutments. It is known as a portal frame type of bridge. Reinforced concrete or steel is used for this type of design, which is applicable to single span bridges over a motorway (see Plate 11).

Where longer spans are required, particularly over rivers, the arch bridge may be a possible choice of design. This may be either a fixed, two-hinged or three-hinged arch. A further type, known as the tied arch or bowstring girder bridge, is an adaptation of the two-hinged arch and resembles an archer's bow because the ends of the arch are tied together horizontally so that there is no horizontal thrust on the abutments.

In fixed arches the abutments are assumed to remain rigidly fixed in position. It is therefore usual for this type to be chosen only where the abutments can be founded on rock or ground with no possibility of settlement (see Plate 5).

In the two-hinged arch, the hinges are placed at the springings and in the three-hinged arch the third hinge is introduced at the crown of the arch. In both cases any slight vertical settlement of any hinge, if moderate, is immaterial. The abutments for arch bridges are quite different in shape from those for other types of bridge, due to the large horizontal forces which have to be dealt with. They are usually constructed in mass concrete and of massive construction to assist in deflecting the thrust into a more vertical direction.

Construction of the arch may be in either steel or reinforced concrete. The space between the top of the arch ribs and the deck is known as the spandrel. This may be kept open and the deck supported by means of columns or transverse walls, in which case the bridge is known as an open-spandrel type. Alternatively, for concrete construction, solid spandrel walls may be built and the space between is either filled or a cellular form of construction is adopted.

The largest spans require the provision of suspension bridges. The principles here are generally well understood. A horizontal deck is suspended from the curved main cables, which are anchored at the ends and pass over the top of a pair of tall piers. The piers may be constructed in steel or concrete, the former being generally preferred at the present time as a very slender section can be used to give a lighter appearance.

Another form of bridge using cables is the cable-stayed bridge, a continuous deck being supported at intermediate points between the piers by inclined cables carried over a tower.

Foundations

An important factor in bridge design is the consideration which must be given to ground conditions at the site. The foundations of the abutments and piers must be capable of transmitting the load to the ground without exceeding the permissible bearing pressures for the material encountered.

If rock or stiff clay or other satisfactory materials are located at, or close to, the proposed level of the underside of the foundations, their design is comparatively easy. Where the depth of such material is not excessive it may be economic to remove the overlying weak material and replace it with mass concrete or extend the abutments and piers to the lower level. Beyond this it is necessary to resort to the use of bearing or friction piles whose main function is to transfer the load from a higher level where the ground may be unable to support it, to a lower level where the bearing capacity is greater. They may penetrate to a firm stratum such as rock, stiff clay or gravel, when they are known as end-bearing piles. Alternatively, they may act as friction piles relying on the friction between their surfaces and the surrounding materials, such as soft clays and silts.

The choice of the type of pile is governed largely by ground conditions. The conventional driven pile is usually of reinforced or pre-stressed concrete, although in special circumstances steel H section, box section or tube is sometimes used. Concrete piles must be designed to be capable of developing the strength necessary to withstand the transporting, handling and driving stresses without damage. These types of pile are driven to a 'set' which is the depth of penetration in inches for each blow of the driving hammer. From this, it is possible to calculate the bearing capacity of the pile.

The other main types are driven or bored cast-in-place piles which rely basically on driving a tube or boring a hole in the ground. The void created is then filled with concrete. Various patented methods are employed with the object of ensuring that the piles are and remain of full cross-section.

Bearings and expansion joints

Apart from the main components of a bridge so far examined the engineer has to consider several other items of equal importance in formulating his design.

For example, the superstructure has to be supported so that the loads are transmitted satisfactorily to the abutments and piers. This is achieved by the provision of bearings which have, in themselves, to be designed to meet the particular circumstances. They must be capable of dealing with all vertical loads as well as horizontal forces due to such things as wind effects, and they must also be able to accommodate any horizontal movement arising from thermal expansion.

For spans of less than 30 feet it is not generally necessary to make provision for expansion, and the bearings may consist merely of providing a bedding for the beams or the slab. For longer spans, however, it is usual to fix the bridge at one end and have the other end free so that it is capable of accommodating the expected movement. Between 30 and 50 feet a simple form of expansion bearing consisting of an arrangement of sliding plates may be satisfactory at the free end. With this type an appreciable amount of frictional resistance can develop, particularly if materials subject to the effects of corrosion are employed. It is, therefore, desirable for the contact faces to be lined with stainless steel, phosphor bronze or plastics to reduce the effect.

Fixed-end bearings may consist of a simple plate to spread the load with a retaining pin or dowel to prevent horizontal movement. It is, however, necessary to allow for angular rotation arising from the deflection of the span. Provision can be made for this by merely crowning the surface of contact to give a point bearing. Alternatively, on the longer spans a fixed rocker may be provided.

A recent development has involved the use of rubber bearings for medium spans. These rely on the ability of a rubber pad carrying a vertical load to move in a horizontal direction and allow for angular rotation.

Roller or rocker bearings made from steel are used for medium spans. These types are also invariably used for long spans and, where large amounts of movement are involved, their design may be quite complicated. In such cases the loads to be carried may be very high and special quality steel will be required.

With multi-span continuous bridges the arrangement of fixed and expansion bearings has to be carefully considered.

It is of course necessary to provide adequate clearance between the deck slab and the abutment, and between adjoining deck slabs to allow for expansion. For short and medium spans a steel plate bedded in bitumen is usually quite adequate for supporting the surfacing material over the joint.

For longer spans, overlapping steel plates exposed on the surface may be used to carry traffic across the joint. Alternatively, where the expected movement is likely to be great, interlocking forked steel plates provide a satisfactory solution and have been used on many bridges.

Surfacing and parapets

The surfacing applied to bridge decks is normally two-course hot rolled asphalt. The thickness is related to that laid on the approaches. For example, with the present standard of a four-inch thickness laid on the carriageways of the motorway, it is preferable for a short or medium span bridge carrying the motorway to have exactly the same. The paving machine can then work continuously over the bridge and thereby produce a better standard of riding quality than if a change of thickness were required. With long bridges, however, the economic considerations of both unnecessary thickness and increased loading may have to be considered.

Unfortunately, hot rolled asphalt is not completely waterproof and it is necessary to provide a waterproof membrane to the deck slab before surfacing. This often consists of a copper-lined bituminous sheeting secured to the deck with a full bed of bitumen and with joints suitably lapped. In order that this should not be damaged during subsequent operations, either by site traffic or by the surfacing equipment, a protective layer of a bituminous sand carpet is usually provided.

Parapets are generally of the open type except in the case of bridges over railways, where statutory requirements demand that solid parapets of a specified height should be provided.

The appearance of bridges

Much attention has been given to producing bridges of good appearance and in conformity with accepted aesthetic principles. The designs of most major bridges are submitted to the Royal Fine Art Commission for approval.

A feature of the motorway bridges in Lancashire is the variety of colour which has been used in the painting of the exposed steelwork.

My interest in these effects was roused in the early 1950's when travelling by train through a Lancashire industrial town. I noticed that the structural steelwork and pipework of a chemical plant in a very drab area of the town had been painted in bright colours. The result so impressed me that I resolved to do all that I could to brighten the appearance of the County. My other concern was to provide something of interest to the driver on the motorway and prevent him from perhaps falling asleep on what could be a long monotonous journey. Much favourable comment has been received from members of the motoring public, not only with regard to the various colours used, but also the variety of designs.

In the treatment of concrete surfaces, it is not easy to achieve a completely uniform smooth texture if the area is large. If the joints between sections of face shutter are not well made, leakage of grout from the concrete can occur and a blemish is left on the surface.

The other problem which arises is in forming what are known as construction joints. These are necessary for various reasons, sometimes associated with the design, but more often at the end of a day's work. It would be preferable for the concrete to be poured continuously but this is generally neither a practical nor economic proposition in the construction of large abutments or wing walls. With smaller structural members, such as columns, there is no reason why this principle should not be adopted because the volume of material is not large and the weight of wet concrete can be reasonably supported over a considerable height.

One method of dealing with these problems is to make a feature of the joints between the shutters in the form of a rebated groove. The construction joints can then be positioned to coincide with the grooves.

A similar principle may be adopted by using tongued and grooved boarding and leaving the surface of the concrete untreated so that a degree of controlled roughness is produced. Construction joints can then be positioned at any joint between the boards and any slight surface blemish would not be obvious as with a smooth finish.

A textured finish can also be produced by tooling the surface of the concrete after it has hardened. Although hiding minor surface blemishes, this may in fact accentuate defects caused by bad joints. Bush hammering, which uses a power tool with a multi-point head, is quite often employed, particularly where this treatment has the effect of exposing coloured aggregates used in the concrete.

Smooth concrete surfaces have been painted, using emulsion paints, with neutral shades of colour to produce pleasing results. Another method is to apply material containing silicone to the surface to prevent the unsatisfactory appearance produced when the surface becomes saturated in wet weather.

Facings of masonry, brick or precast concrete blockwork probably provide the best solution but involve extra expense. They enable variety to be introduced into bridges of a standard basic design. In theory it is possible to use the facework as a shutter and cast the concrete behind, but this leads to a number of problems. Firstly, it restricts the height of each lift of concrete to the height of facework wall which is self-supporting, unless additional support is provided. Secondly, any concrete spillage from the upper lifts of concrete will affect the appearance of the lower facework unless precautions are taken. It has been found in practice that it is preferable to cast the concrete first and then build the face work as a last operation.

Because of the high cost of natural stone and the shortage of the skilled labour required to prepare and lay it, use of this material can only be justified in exceptional circumstances. Brickwork lends itself to the more urban areas, where it generally harmonises with existing surroundings, but in open country the rough texture capable of being produced in concrete blockwork will be more suitable.

Motorway bridges in Lancashire

The total number of bridges constructed on the 61½ miles of M.6 in Lancashire was 174. With the 22 bridges on the Stretford-Eccles By-pass, M.62, this meant that a total of 196 motorway bridges had been designed and the construction supervised by my Department in the first ten years of motorway construction in Britain. A further 290 bridges are under construction or being designed, thereby increasing the total number to 486. Of these, 190 are or will be of steel girder construction with reinforced concrete deck slabs involving approximately 90,000 tons of bridge steelwork. The remaining 296 bridges are or will be of reinforced and pre-stressed concrete.

Considerable experience has been gained from the many different types built with these construction materials. In many cases the economic and other advantages in the use of one material as compared with another, are very finely balanced.

The present state of the economy demands that the economics of alternative types are fully investigated. Many factors enter into these considerations and affect the choice of bridge type and construction materials. The nature of the site; the ground conditions, including the possibility of mining subsidence; whether the crossing is square or skew; road alignment in so far as visibility may be affected; erection problems in relation to access to the site; time available for both design and construction, particularly in respect of advance bridgeworks: all these must be taken into account.

Three examples of those already built or under construction are described below and others are illustrated.

Barton high-level bridge (Plate 3) carries the Stretford-Eccles By-pass, M.62, over the Manchester Ship Canal. It is 2,425 feet long and rises to a height of some 100 feet above water level with a maximum gradient of 1 in 25.

The piers consist of two-leg portals of reinforced concrete construction on piled foundations and vary in height from 30 to 80 feet above ground level. The columns are hollow and fitted with permanent steel ladders which provide easy access from the ground to the tops of the piers for inspection and maintenance.

The superstructure consists of eight steel plate girders carrying a reinforced concrete deck slab in two halves, one half for each carriageway. The total width is 73 feet incorporating dual 24 feet wide carriageways, a central reserve and side verges. No hard shoulder is provided because of the extra cost involved in the extra width.

There are eighteen spans in all, of which thirteen of 115 feet each and two of 135 feet each are simply supported. The two anchor arm spans to the canal crossing are 175 feet each. The main span is 310 feet and consists of two cantilevers each 77 feet 6 inches long carrying a simply supported 155 feet centre suspended span.

The girders, which are all of riveted construction, are generally 9 feet deep, except over the main piers to the canal crossing, where the depth is increased to 18 feet.

The treatment of the steel was carried out on the site in shops erected for the purpose. After shot blasting and applying the first coat of priming paint to sections of the girders, they were generally spliced on the ground prior to erection.

The approach viaducts were erected by mobile cranes. The main span, and the anchor arms and cantilevers, were erected by two different methods. In the first, the anchor arm and cantilever girders were lifted as complete units by means of a frame mounted on top of the main pier and a large derrick mounted on a rail track, both working together. In the second, the pieces were lifted individually and supported on temporary trestles while the riveting of the splices was carried out above ground.

The suspended span for the western half of the bridge was launched by means of a Bailey bridge launching nose and then jacked down into position. After concreting this half of the deck, it was used to position the girders for the other half of the span. These were assembled on bogies on rail track laid on the approaches, moved up into position and lifted sideways by two cranes on to their bearings.

The bridge came into use when the By-pass was opened to traffic in October 1960.

The Thelwall bridge (Plate 4) is 4,417 feet long and, when opened to traffic in July 1963, was the longest bridge on a British motorway. It carries the M.6 Motorway over the Manchester Ship Canal and the River Mersey, rising to a height of 93 feet above the water level in the canal with a maximum gradient of 1 in 25.

There are 36 spans, 30 of which are 110 feet in length. The three spans at the canal crossing are 180 feet, 336 feet and 180 feet long. The main span has a centre suspended span of 168 feet. At the river crossing there are three spans of 118 feet 6 inches, 180 feet and 118 feet 6 inches in length.

In view of the length of the bridge it was considered desirable to limit the number of joints in the deck to a practicable minimum so that the riding quality of the surface would not be impaired. The deck on the approach spans was, therefore, made continuous in three sections, each of ten spans.

The bridge is 92 feet wide between parapets to accommodate dual 36 feet wide carriageways, a central reserve and side verges. As in the case of Barton high-level bridge, hard shoulders are not provided.

The terrain consists of alluvial deposits and over the area between the canal and the river there is an overlying deposit of silt from canal dredgings, some 45 feet deep.

On the south side of the canal and also at the canal crossing, the foundations to the piers are carried on piles driven into firm sand or sand rock underlying the clay. On the north side of the canal, in the deposit area, piles were driven into a compact bed of sand and gravel overlying the clay. Two types of pile were used. Shell piles 17½ inches and 20 inches in diameter, and carrying working loads of 75 and 100 tons respectively, were used for lengths of between 25 and 50 feet below ground level. Cylinder piles between 50 and 130 feet long were also used. The working load is 300 tons when founded on firm sand or rock. In the deposit area, however, the loading is reduced to 150 tons to give a better load spread and to minimise the possible effects of settlement in the clay below the sand and gravel bed. The cylinder piles were made up of spun concrete tubes with a wall thickness of five inches and an external diameter of 36 inches. The tubes were each 16 feet long and holed longitudinally to take stressing cables. After assembling a number of tubes to give the required length, they were stressed together and the cable ducts were grouted.

The piles were pitched and inserted into holes drilled through the upper strata down to the founding level of the hard sand. They were then driven to the required set using a ten-ton hammer and filled with concrete.

The piers generally consist of four-legged reinforced concrete portals varying in height from 30 feet to 80 feet above ground. The canal crossing piers are of hollow wall construction, fitted with ladders to give access to the superstructure, and those of the river crossing, which are relatively short, are solid reinforced concrete walls.

Apart from the River Mersey crossing, which is widened on the west side to accommodate a deceleration lane to the adjoining interchange, the reinforced concrete deck slab is supported on eight steel plate girders, four to each carriageway. The canal crossing girders are of riveted construction and vary in depth from 10 to 20 feet. The remaining girders are of welded construction eight feet six inches deep throughout, except at the river crossing, where the depth is increased to 16 feet.

The suspended span on the canal crossing was erected by cantilevering out in sections from one of the cantilever arms for two-thirds of its length. The remaining one-third was erected from the other end and, after jointing, jacked down into position. Expansion movement of up to seven inches has to be catered for and a forked type joint of cast steel has been provided.

Because of the magnitude of the project, an advance contract was let ahead of the adjoining length of motorway and work started in September 1959.

At the same time work also began on Gathurst viaduct, another major bridge for M.6, again in advance of the motorway construction.

The Lune bridge (Plate 5) carries the Lancaster By-pass section of M.6 over the River Lune.

It is a reinforced concrete open-spandrel fixed-arch with a clear span of 230 feet and a rise of 44 feet. The overall length is 400 feet and the width between parapets is 97 feet. It incorporates two separate arches of cellular construction three feet six inches deep at the crown and 6 feet deep at the springing. The spandrel walls support a reinforced concrete deck.

An interesting constructional feature carried out by the contractor involved the use of the same scaffolding for both halves of the bridge without dismantling. A timber gantry built across the river supported the scaffolding and shuttering for concreting the first arch. This was later lowered slightly, winched sideways on to a second gantry as a complete unit, raised to its correct level and used to form the second arch.

The elevation face of the arch is bush-hammered. The spandrel walls and deck structure, however, have a smooth finish to the concrete and the soffit of the arch is left untreated with the board marks of the shutters left untouched.

A feature of the bridge is the echo produced by the arch. From a point on the edge of the river near the springing of the arch any sound is reflected back upwards of twelve times.

The bridge was opened to traffic when Lancaster By-pass came into use in April 1960.

Future trends in motorway bridge design and construction

Great emphasis is placed, at the present time, on the need for economy in the building of motorways. With regard to motorway bridges several suggestions have been made as follow:

1. More standardisation of units.
2. An increase in the amount of off-the-site prefabrication of components.
3. The greater use of a centre support in the case of overbridges. The effect of this would be for overbridges to be of two or four spans only.
4. The use of new and improved construction materials.
5. The elimination of decorative treatments such as facework.

It is true to say that all or any of these principles may not always achieve the desired object and each design problem must be treated on its merits. Decorative treatment should not always be deleted if the extra cost is small.

Chapter 9

Contract procedure

Although some construction work has been carried out by the direct labour organisation of local highway authorities, the vast majority of motorway construction has been undertaken by contract.

The type of contract normally used is based on the following documents:

(a) Conditions of Contract
(d) Drawings
(c) Specification
(d) Bill of Quantities

Conditions of Contract

The conditions applicable to a particular contract are framed on a basis of the 'General Conditions of Contract for use in connection with Works of Civil Engineering Construction',[1] commonly known as the I.C.E. Conditions, because they were prepared jointly by the Institution of Civil Engineers, the Association of Consulting Engineers and the Federation of Civil Engineering Contractors, and are recommended for use by those bodies.

There are normally two parties to the contract, the 'Employer' and the 'Contractor', and in the case of motorway construction the 'Employer' is the Minister of Transport or a Local Authority acting either independently or as an Agent for the Minister. The employer is required to delegate certain powers and duties to an agent known as the 'Engineer' to the contract, who, in the case of a local authority, would normally be a chief officer of that authority.

[1] Institution of Civil Engineers, Association of Consulting Engineers, Federation of Civil Engineering Contractors, *General Conditions of Contract and Forms of Tender, Agreement and Bond for use in connection with Works of Civil Engineering Construction*, 4th edn. 1955.

As County Surveyor and Bridgemaster of Lancashire, I have acted in this capacity for motorway contracts carried out over a period of twelve years in the County.

A consulting engineer in private practice may, however, be appointed by an employer and act as engineer for the purposes of the contract.

It is usual for the engineer to the contract to have previously been engaged in the design of the scheme and in the preparation of the drawings and documents concerned. In this capacity he is invariably required, at the various stages, to provide estimates of cost to the employer. Considerable skill and experience is required in this aspect of his work, in order that the employer may be aware of the financial commitments at all times.

The engineer has many responsibilities under the contract. Not only must he see that the employer's interests are protected, but he must also act impartially in the event of disputes arising between the employer and the contractor. If either the employer or the contractor is dissatisfied with any decision of the engineer, the matter may be referred to an independent arbitrator to be agreed between the parties, or failing agreement, to be nominated by the President of the Institution of Civil Engineers.

It is usual for an engineer's representative to be appointed to supervise the actual construction on behalf of the engineer and the employer. On the large motorway contracts he is known as the Resident Engineer. In addition to examining and testing the materials to be used and the workmanship employed in connection with the contract the resident engineer may have delegated to him any of the powers and duties vested in the engineer.

The resident engineer may be an assistant on the staff of the engineer and may have been involved in the design and the preparation of the contract documents. Alternatively, he may be appointed merely for the period of the contract either by the engineer or the employer.

The responsibility for any actions of the resident engineer remains with the engineer and the employer. Therefore, the decisions which he makes should be those which he would expect the engineer to make in similar circumstances. A clear understanding should be developed between the two, so that the resident engineer may be free to use his initiative, with confidence existing on both sides. The advantages of the resident engineer being a member of the engineer's staff are obvious.

The contractor is also required to provide a representative, approved of in writing by the engineer, for the full-time superintendence of the work and he is known as the Agent. On a large scale motorway contract he will usually be a qualified civil engineer, similar in status to the resident engineer.

Sub-contractors

The contractor is not allowed to sub-let any part of the work without the written consent of the employer. However, if the suggested sub-contractor is well-experienced and has a good reputation for the type of work to be sub-let to him, then the engineer's consent is usually given. It is usual for contractors to be required to list their proposed sub-contractors in their Tender so as to give some indication of the work to be undertaken by this method.

Contractor's programme

As soon as practicable after the acceptance of his tender, the contractor, if required, must submit to the engineer for his approval a programme showing the procedure he proposes to use for carrying out the work. This is extremely important, particularly if the motorway contract concerned is related to any other works which are to be co-ordinated with it. It is possible, of course, to insert special conditions in any contract to require certain operations to be completed to a particular time schedule. This, however, restricts the contractor's freedom of action in carrying out the work as he thinks best, depending on the resources at his disposal. Restrictions of this kind may lead to increased cost, so that they should only be stipulated when absolutely necessary.

In view of the complexity of modern construction techniques and increasing mechanisation, careful phasing of all operations in the construction of a motorway is vital. A single item of work can seriously disrupt progress by upsetting the proper sequence. Many contractors now use the Critical Path Method for planning and programming the construction work on motorway projects. The *Second Report on Efficiency in Road Construction*[1] recommends that the contractor should be required to submit a CPM Network for major works soon after he is appointed.

Setting-out

The contractor is responsible for the proper setting-out of the work. The design of the project will have been based on certain basic information obtained from the initial survey. On the completion of the design, it is necessary for sufficient data to be made available in the contract documents to allow the contractor to set-out the work accurately. These data are related to physical features on the site and, where necessary, to markers established by the engineer for the purpose.

On a large project extending over many miles of open country, considerable skill has to be exercised in carrying out this task. It has become the practice for the

[1] National Economic Development Office, *Second Report on Efficiency in Road Construction*, London, Her Majesty's Stationery Office, 1967.

contractor to employ specialist firms for the purpose, equipped with the most up-to-date surveying equipment. For example, instruments which use electronic methods are particularly applicable for measuring long distances across uneven ground, rather than the simple chain or steel tape.

Accuracy is extremely important. Where bridges and other structures have to be constructed on isolated sites in advance of other works, these must be sited with absolute precision. Land boundaries must be accurately defined and comply with the land plans.

Setting-out is a continuing process and must always be ahead of the various stages of the work to which it applies. The pegging-out provided on the ground for earth-works will, in due course, have to be followed by sight rails applicable to the drainage works. Later, the carriageway works will be required to be set out.

Similarly, in the case of bridges, the setting-out for the foundations will be followed by the accurate location of the abutments and piers. The positioning of bearings for bridge girders requires high precision in order to ensure that the factory-made components can be erected without difficulty.

Care of the works and insurances

Responsibility for the safety of the works, for watching and lighting, and for the safety and convenience of the public and other third parties, lies with the contractor. Because of this he is required to take out insurances applicable to the contract specifying the minimum amount of third party cover required.

Temporary works

Unless anything is specified to the contrary, the methods employed by the contractor for carrying out the work are entirely his responsibility.

When required by the engineer, the contractor must submit details of his proposals for approval. The conditions of contract, however, expressly state that the submission or approval does not relieve the contractor of his responsibilities.

The engineer is clearly involved if the methods proposed are likely to affect the finished work in any way. For example, the method of erection proposed for a bridge might be such that it would overstress the members. In this case the engineer would be quite justified in refusing to approve the method and would be able to prove his point by calculation.

The contractor is required to comply with the Construction Regulations under the Factories Acts, which provide a code of safety legislation specifically designed to apply to works of civil engineering construction. These regulations also require the contractor on works of any size to nominate a suitably qualified Safety Officer responsible for safety supervision.

Variations

The engineer is empowered to vary the work as he considers necessary, and the contractor must accept his instructions to this effect. The amount to be added or deducted from the tender sum consequent upon the variation is determined by the engineer. In any motorway contract it is impossible to allow for every eventuality which might arise and some variations are inevitably necessary. The financial implications, however, may be considerably in excess of the value of the items of work varied. This is due to the effect on other operations which might arise. It is unfortunately true that apparent savings on one particular item may increase the overall cost!

Circumstances may arise where physical conditions or artificial obstructions are encountered which could not have been reasonably foreseen by an experienced contractor. A procedure is laid down which covers the action to be taken by the contractor so that an opportunity is given for the engineer to agree the methods to be adopted for overcoming the problem.

Certificates and payment

After the end of each month the contractor is required to submit to the engineer a statement showing the estimated contract value of the permanent works executed up to the end of the month. If the value is sufficiently large to justify the issue of an interim certificate, then payment will be made. It is usual for a minimum amount of an interim certificate to be specified in the contract. For large motorway contracts this would be around £20,000.

Period of maintenance

The contractor is responsible for carrying out maintenance and making good any defects for a specified period after completion, usually twelve months. After this, the work is handed over to the employer in a condition satisfactory to the engineer and the other half of the retention money is payable. The contract is then considered to have been completed.

Other matters dealt with in the conditions of contract include the procedure for dealing with claims, the settlement of disputes by arbitration, and such eventualities as the contractor becoming bankrupt.

Drawings

The number of drawings required for a single motorway contract may amount to many hundreds, as indicated in Table 5 in Chapter 4. They should be capable of indicating to the contractor the engineer's requirements in order that, firstly, he can price his tender and secondly, carry out the work.

The scale of the drawing depends on the size of the feature to be described. For a motorway roadworks plan in a rural area, a scale of 1/1250 was formerly considered to be adequate. It has, however, been found desirable to use a larger scale and 1/500 is now generally accepted as being necessary.

In addition to the plan, drawings showing the longitudinal and cross-sections are required. The scale may be 'natural', which means that the horizontal and vertical scales are the same. Alternatively, the vertical scale may be larger than the horizontal scale, which will thus show clearly any minor variations in level. In such cases the section will be described as having an 'exaggerated vertical scale'. The main purpose of sections is to give levels for setting-out the work, and for computing quantities.

The importance of the drawings being accurately and carefully prepared cannot be over-stressed. Errors and omissions which become apparent only as the work is being carried out, can lead to delay and disruption, with a resulting increase in cost.

Specification

The Specification gives a detailed description of the nature and quality of the material and workmanship to be used in carrying out the work. For motorway contracts the Ministry of Transport *Specification for Road and Bridge Works*,[1] modified and supplemented by additional clauses, is used. This document makes use of many of the standards of the British Standards Institution. These are developed by Committees consisting of representatives from all interested parties and consequently cover standards of workmanship which can be reasonably achieved and materials which industry is equipped to produce economically. In addition, standards developed by Trade Associations are quite often incorporated in the specification.

General

The provision of site offices for the resident engineer and staff, the control of traffic and the building of temporary haul roads for access to the site, are matters covered by the general series of clauses. The contractor is responsible for these and is allowed to price the requirements in the Bill of Quantities on a lump sum basis.

[1] Ministry of Transport, *Specification for Road and Bridge Works*, London. Her Majesty's Stationery Office, 1963. (Superseded by a revised edition introduced in April 1969.)

A further series of clauses deals with the standard of workmanship required of the contractor and his general conduct in the execution of various aspects of the work. An example of this is the restriction imposed in cold weather conditions where concrete would be adversely affected by frost.

The Specification then describes the work to be carried out.

Site clearance

The first operation on any motorway contract is the clearance of the site which involves the removal of buildings, walls and other obstructions of a similar nature.

Fencing

Permanent motorway fencing generally consists of a wooden post and four-rail fence of the nailed type, conforming with the appropriate British Standard. Where additional stock-proofing is required, the gap between the rails is reduced by either adding a fifth rail or providing horizontal wires between the rails. The permitted types of timber are defined and the preservative treatment is prescribed for each of the types. Other kinds of fencing, the provision of gates and stiles, etc. are also described, including safety fences which are a well-known feature of motorways.

Drainage

Of the many different types of pipe suitable for use in motorway drainage systems, glazed clay and concrete are probably the most generally used, but pipes made of materials such as asbestos cement, corrugated metal, pitch fibre and plastic (P.V.C.) have been widely employed. Due to their higher cost the use of cast iron and steel pipes is usually confined to special cases where special constructional problems exist.

The principal recent advances in the manufacture of pipes have been improvement in strength, and in new jointing methods. Many different types of joint are in use. What is required is a joint which is efficient and which can be quickly made, thus reducing labour time on the site. Most modern types of joint employ the use of rubber rings which provide a good seal and allow the pipeline to flex. This is especially important in conditions where settlement is likely. With pipes made of materials such as corrugated metal or P.V.C., which are themselves flexible, the joint is usually made by means of a sleeve.

The strength of a pipe in a pipeline is related to the condition of the bed upon which it rests and the material backfilled around and immediately above it. It is therefore important that the contractor excavates the trench in a controlled fashion and, where necessary, supports its sides. The specification must also detail the requirements for bedding and backfilling, depending on whether rigid or flexible pipes are to be used.

For French drains the pipes may be of porous material such as 'no-fines' concrete, or clay tiles with open joints, or alternatively any type of perforated pipe. For a drain of this type to function properly, the filter material used in backfilling the trench must be carefully graded, depending on the surrounding ground. It is required to be hard clean crushed rock, slag or gravel. For normal use the size is graded from $2\frac{1}{2}$ inches to $\frac{3}{8}$ inch. In silty ground, however, it is important to ensure that the silt does not find its way through the filter material and block the pipe either externally or internally. In such circumstances a finer grading is used, varying from $\frac{3}{8}$ inch to material of a smaller particle size.

Earthworks

Earthworks and the problems associated with them generally present the greatest difficulties in carrying out any motorway project in Britain and this is particularly true of the conditions encountered in Lancashire.

In specifying the earthworks requirements in a contract, it is essential to try and adhere to the following basic principles:

1. The completed earthworks should be a sound engineering job.
2. The design and specification should be economic.
3. The information made available to the contractor should be such as to allow him to tender on a fair and reasonable basis.

The engineer in deciding on the profile for the motorway, attempts to balance the volume of excavated material which is suitable for use in the embankments with the amount of material required for the embankments. In other words, no material has to be brought on to the site from outside and only unsuitable excavation has to be removed. Although this is the ideal aimed for, it is rarely achieved. The practical and economic limitations imposed on the soil survey clearly do not permit the estimation of volumes of suitable and unsuitable material with 100 per cent accuracy.

The following major earthworks operations are likely to be involved:

1. The excavation of suitable material and the haul, deposition and compaction of this material in embankments.
2. The excavation of unsuitable and surplus suitable material and its haul and deposition on to tips off the site.
3. In the event of a deficiency of suitable material, the supply, deposition and compaction of suitable imported filling material.

One of the main requirements of the earthworks specification is thus to define which materials are suitable and which are not. The easily recognised unsuitable materials do not present any difficulties, and include those excavated from swamps, marshes or bogs and running silt, peat, logs, stumps, perishable material, domestic refuse, and highly combustible material. At the other end of the scale, gravels, sands and rock broken down to a satisfactory size are quite suitable.

The difficulties which arise are almost entirely concerned with the plastic or cohesive soils, such as clays, where the moisture content is highly critical in relation to its mechanical properties. With these materials it is possible to define their suitability or unsuitability only by referring to various terms associated with the science of soil mechanics. For example, the 'plastic limit' of a soil is the moisture content at which it ceases to be in a plastic condition as determined by a specified test carried out on a sample. Similarly, the 'liquid limit' is the moisture content at which the soil passes from the plastic to the liquid state. The 'plasticity index' is the numerical difference between the liquid limit and the plastic limit.

Clays with a very high liquid limit coupled with a very high plasticity index are unsuitable in any event, as their mechanical performance is unsatisfactory.

With other clays it has been the practice to declare as unsuitable any material which on excavation has a moisture content higher than a specified margin above the plastic limit. A figure of 2 per cent has been used in recent specifications, but it has been found that satisfactory results have been possible with certain clays at much higher moisture content. Where, however, a contractor has tendered on a basis of working to a clearly defined specification such as that quoted, and it has been found necessary to vary this during the course of carrying out the work, contractual difficulties may arise.

The responsibility for finding tipping sites lies with the contractor unless specific requirements to the contrary are defined in the contract. The contractor will thus need to negotiate for the necessary land and obtain planning permission under the Planning Acts. A similar procedure has also to be adopted for opening borrow pits to obtain imported filling.

When tendering for the contract the contractor will most likely prepare a form of mass haul diagram which will indicate for each cutting how the excavation is to be disposed of by reference to the particular embankment or tip concerned. Likewise, the source of materials, excavated and imported, will be shown for each embankment. It will be realised that the interpretation of the initial soil survey is important not only in the design and preparation of the contract documents, but also in the formulation of the tender.

The final cost of the scheme will be affected by the amount of imported material. The intention of the Ministry of Transport's 1969 revision of the specification is to give the contractor a positive incentive to use as much excavated material as possible.

The specification gives details of how the embankments are to be formed and defines the thickness of layer to be laid at one time for various materials. For example, rock used in rock-fill embankments has to be of a size that can be deposited in horizontal layers not exceeding 18 inches loose depth and spread and levelled by means of a crawler tractor weighing not less than 15 tons. Isolated boulders can be used, however, in the body of embankments not of rock fill, providing they are not placed within 2 feet of the formation level.

It is most important that the embankments are built up evenly over the full width and at all times to have an even surface and sufficient crossfall to enable rain water to drain away.

The contractor is required to carry out compaction trials for each type of material to be used in the embankment, before starting work with that material. The object is to demonstrate that the type of compaction plant and number of passes made by the plant is capable of producing the specified compaction requirements.

Control of compaction in embankments is necessary to eliminate subsequent settlement. The specification requires that the material after compaction shall have a dry density corresponding to not more than ten per cent air voids at the moisture content concerned, except for the top two feet of filling where the air voids shall be not more than five per cent.

Reference has already been made to the limiting moisture content for plastic materials. In the case of non-plastic materials it is limited to the optimum moisture content and three per cent below. The optimum moisture content is that at which the most dense condition can be achieved.

Once it has been established that a certain type of plant operated in a particular manner is capable of producing the desired results this can be used as a means for controlling the operation, subject to carrying out periodic density tests as a check. There are practical difficulties in carrying out density tests in rock fill. The method of compaction is, therefore, defined by giving the number of passes for various types of roller or other compaction equipment.

The surface soil is required for use in the soiling of slopes of cuttings and embankments, verges and central reserves. It needs to be stockpiled either inside or outside the site for use as and when required.

One of the most important aspects of earthworks operations involves the preparation of a satisfactory formation, which is the surface of the ground in its final shape after the completion of the earthworks. To avoid damage by construction traffic, a protective layer of material is left in cuttings and an overfill provided on embankments; this material is removed only in the final trimming operation after the drains have been laid and immediately before the carriageway construction is due to start.

It is important that the formation in cuttings is protected immediately after it has been exposed. To maintain the moisture content at its natural level, one method sometimes adopted is to surface dress the formation with hot tar or bitumen and blind with gravel, crushed rock or slag of a $\frac{3}{16}$-inch maximum size. An alternative is to lay polythene sheeting. It is more economical, however, to lay the sub-base immediately behind the trimming operation.

It is not uncommon to encounter pockets of material in the formation of a cutting which are quite unacceptable and on which it would be undesirable to lay the carriageway construction. In such cases the material may be excavated and replaced with selected excavated maetrial or imported granular material compacted similarly to that used in embankment construction. Occasions may arise where the backfill has to be deposited below standing water. In such instances, compaction is impractical and a suitable free draining material is used which does not require compaction under such circumstances.

An alternative method of treating the formation is to place rock fill varying in size from 15 to 6 inches and punch it into the surface using heavy compaction equipment such as rollers. This is continued until a stable condition is reached. Any poor material which has been forced to the surface is then removed.

Sub-bases

Sub-base materials must be capable of meeting the requirements of Road Note 29[1] in which two gradings are specified: Type 1 which has a maximum size of $1\frac{1}{2}$ inches and must be either crushed rock, concrete or slag, or well burnt non-plastic shale; and Type 2 which is much smaller and finer and of the same materials or of well graded sands and gravels. Type 1 is intended for use where site conditions are likely to be wet or where the sub-base will have to carry construction traffic.

It is common practice for specifications to be prepared on a basis of proved materials normally available in the locality. For example, in Lancashire crusher-run limestone is readily available and specifications have been framed accordingly. It has been the practice on recent motorway contracts to use a six or seven-inch thickness as the top layer of the sub-base. In this instance the maximum size of the stone is three inches. Crusher-run limestone is usually comparatively coarsely graded and it is then necessary to add fines of a size from $\frac{3}{16}$ inch to dust in order to bind the material together and tighten the surface. Other materials are used in the lower layers.

Another type of sub-base occasionally used is soil-cement. This process involves the stabilisation of a layer of naturally occurring or processed granular materials, or

[1] Ministry of Transport, *A Guide to the Structural Design of Flexible and Rigid Pavements for New Roads*, Road Note 29, Second Edition, London, Her Majesty's Stationery Office, 1965.

a combination of both, by means of cement to provide a material of specified strength. The operation is carried out on the soil using plant capable of pulverising and fully mixing the materials together.

Bases

The materials acceptable for use in the base for motorways are lean concrete, dense tarmacadam, dense bitumen macadam and rolled asphalt.

Lean concrete may have aggregate with a maximum size of either $1\frac{1}{2}$ inches or $\frac{3}{4}$ inch. The ratio of cement to aggregate by weight has to be between 1 : 15 and 1 : 20. The contractor is required to produce material of a specified strength and achieve a minimum density after laying and compaction. The design of the mix is most important, bearing in mind that the strength varies with the water/cement ratio, the lower this ratio the greater the strength.

Dense tarmacadam and dense bitumen macadam are very similar materials, except that the binder in one case is tar and in the other bitumen. The two types of binder have different characteristics and so the amount required varies. The size of the aggregate used is related to the thickness of the layer to be laid and this again affects the binder content. Materials are manufactured in $1\frac{1}{2}$-inch and one inch nominal sizes, and the binder content varies from 3·5 to 4·7 per cent by weight of the mixture for crushed rock aggregates. The specification covers binders of differing viscosities and the engineer must specify the viscosity required to suit the particular circumstances. In the case of bases where strength is required it is usual to specify harder binders.

Rolled asphalt for road bases is required to comply with the appropriate British Standard for the base course of asphalt surfacing and is, in fact, a similar material. The thicknesses of this and other layers of construction are given under 'Types of pavement', Chapter 6.

Surfacing

In the case of flexible construction, the carriageways of the motorway are surfaced with four inches of rolled asphalt laid in two courses. This is a hot-processed material consisting of a mixture of a relatively hard bitumen with aggregates, in such gradings and proportions that when hot it can be spread and compacted with a roller. The materials are manufactured in accordance with B.S.594 Rolled Asphalt (Hot Process).

For the base course, the coarse aggregate content is 65 per cent. The binder, or asphaltic cement, as it is known, is petroleum bitumen derived from oil refining. A low stone content material is used in the wearing course and the coarse aggregate content is restricted to 30 per cent. The asphaltic cement may be either equal proportions by weight of petroleum bitumen and refined lake asphalt or a pitch/bitumen mixture. In the latter case the pitch is a coal-tar pitch produced in a distillation process and the bitumen is petroleum bitumen. The proportions used are of the order of 25 per cent pitch to 75 per cent bitumen.

To give added skid resistance $\frac{3}{4}$-inch or $\frac{1}{2}$-inch chippings are rolled into the surface. The characteristics of the aggregate used for this purpose and also for the coarse aggregate in the wearing course are, therefore, equally important.

Several different types of test are normally carried out to determine the suitability of aggregates. The one dealing with skid resistance is a test carried out in accordance with British Standard 812 to determine the polished stone value. This measures the extent to which the exposed aggregate is susceptible to polish under traffic. The minimum specified value for the chippings is 59 and that for the coarse aggregate in the wearing course 45.

In the Service Areas of motorways, where there is a considerable volume of standing vehicles, dense tar surfacing is used. Tar is more resistant to the effects of oil droppings. There is no British Standard for this material and the specification used follows the recommendations published by the British Road Tar Association.

Dense bitumen macadam or dense tarmacadam base course material is used for surfacing the hard shoulders. For a two-inch compacted thickness, $\frac{3}{4}$-inch nominal size material is used. The surface is then sealed with slurry seal or a similar approved material. Slurry seal is a mixture of crushed igneous rock or limestone fine material of $\frac{1}{8}$-inch maximum size and a quick-curing bitumen emulsion. As the emulsion breaks, the water evaporates leaving the coated aggregate on the surface. The thickness of the seal is about $\frac{1}{8}$-inch. Laying is carried out by mechanical means and the surface is rolled by means of a self-propelled or towed multi-wheeled rubber-tyred roller, as soon as the slurry has set sufficiently.

Schlamme, already mentioned, is similar in some respects to slurry seal but the size of the aggregate is smaller, and usually consists of sand. It is applied evenly over the surface by brush and generally in two applications.

Concrete carriageway construction

The concrete used in the construction of a concrete carriageway is specified on the basis of a minimum strength requirement, subject to a number of restrictions in respect of size and type of aggregate, a minimum cement content and a maximum water/cement ratio. Within these limitations the contractor is allowed freedom to design the mix to suit the aggregate available and the plant to be used.

The strength of concrete has in the past been based on the crushing strength of cubes cast in moulds of a standard size, usually six cubic inches, the minimum compressive strength required at the age of 28 days being 4,000 lb. per square inch. It has been established that in a concrete pavement the tensile strength of the concrete is more important than the compressive strength, and consideration has been given to framing the specification on a basis of indirect tensile strength or flexural strength.

Where a reinforced pavement is being constructed, the slab is usually laid in two layers. The nominal maximum size of the aggregate in the bottom layer may be $1\frac{1}{2}$ inches and in the top layer $\frac{3}{4}$-inch to assist in obtaining a better finish to the surface. The minimum cement content has generally been based on the aggregate/cement ratio, which should not exceed 7 : 1 by weight. The current practice is to specify the minimum weight of cement per cubic yard of concrete. By specifying a maximum water/cement ratio, usually 0·55 by weight, it is possible for a measure of control to be exercised on a day-to-day basis.

The contractor is required to carry out trial mixes and lay trial lengths of slab to show that the mix proposed and the methods of laying are capable of achieving the desired results.

In the production of concrete, it is important that it should have a minimum variation in strength and consistency. Practical difficulties do not permit absolute uniformity and it is necessary to aim for a higher strength in order that the minimum should not fall below that specified. Statistical methods are used to arrive at the target figure.

Frost and the use of salt to assist in providing safe driving conditions can be detrimental to the surface of a concrete pavement. Modern methods of reducing the effect involve the addition of what are known as air-entraining agents in the top layer of the pavement. These agents produce microscopic air bubbles in the concrete, which resist the effects of a sudden lowering in temperature. The specification requires the contractor to provide a quantity of air equivalent to between three and six per cent of the mix. This is measured by an air meter on the mixed concrete before placing. Adding air can reduce the strength, but improve the workability. The contractor has therefore to take great care in the design of a mix of this type.

Before laying the concrete a waterproof underlay is placed on the surface of the base, polythene sheeting often being used. The reasons for this treatment are twofold. If the base is open in texture, the matrix from the concrete will tend to be driven into it during placing and compaction. If, however, the base is saturated with water, this can be drawn up into the concrete, increase the water/cement ratio and thereby reduce the strength.

After the surface of the concrete has been finished it is brushed in a transverse direction, using a wire broom. A minimum standard of texture depth is specified in order to provide a satisfactory skid-resistant surface. This is measured by a 'sand patch' test in which a known small volume of fine sand is spread over the surface, thereby occupying the grooves formed by brushing. The area of spread is divided into the volume and represents the texture depth. Skid resistance is also ensured by specifying the characteristics of the aggregate used in the top layer of concrete.

It is important that concrete should be protected and cured in its early life. Protection from the rain and hot sun during the first few hours after laying is effected by the provision of tents. The loss of moisture from the concrete over the first few days can be very detrimental; to prevent this a metallised resinous curing compound, which provides a seal, is sprayed on the surface as soon as possible after brushing.

Minimum thickness and surface regularity of pavement courses

It is important that, where a minimum thickness of a particular course is specified, this should be obtained. As the more expensive materials are invariably in the layers nearer to the surface, the contractor will always endeavour to finish each course as high as possible. This requires considerable skill, particularly where the level of the finished surface has to correspond with that of previously laid fixed objects, such as concrete marginal strips.

With the elimination of these features and greater mechanisation in laying, it has been proposed that there should be a new approach to the specification which, while requiring a specified thickness of course to be laid, would not be restrictive on the ultimate level of the surface.

The basis of specifying the surface regularity is defined by the maximum permitted vertical depression on a rolling straight edge, ten feet long, operated parallel to the centre line of the road. For wearing course asphalt and the finished surface of a concrete pavement, this is $\frac{1}{8}$ inch.

Surface regularity is also important in the lower layers of the construction and figures are quoted for each course. For example, in base course surfacing, the vertical depression should not exceed $\frac{1}{4}$ inch under a ten feet long straight edge.

Structures–bridges, culverts and retaining walls, etc.

In work of this kind it is possible to indicate and describe many of the specification requirements on the drawings and by means of typical details.

The major requirements covered by the specification are as follows:

Formwork. A high standard in the construction of the formwork is essential in order to produce satisfactory concrete work. Drawings showing details of the formwork proposed are required to be submitted to the engineer for approval. The standard should be such that the resulting cast work is to an accuracy of $\frac{1}{8}$-inch on a ten feet long straight edge.

Concrete. Several different classes of concrete are specified. The mix proportions are based on the number of cubic feet of the different types of aggregate for one cwt. of cement. These are converted into weights, based on the weight per cubic foot of each type of aggregate which is to be used. The proportioning of the mix is then carried out by means of a weigh-batcher. A minimum compressive strength is also specified for each class of concrete by reference to the cube strength.

It is necessary to carry out trial mixes before concreting is started, in order to establish that the strength can be obtained for the mix and the water/cement ratio proposed. Once the correct consistency has been determined by testing, and approved by the engineer, it must be maintained within specified limits.

Conditions are imposed on the methods of transporting and placing of concrete. Any concrete which has not been placed within 30 minutes after mixing must not be used without approval. The use of purpose-made agitators generally permits an extension of this time to not more than 1½ hours.

Segregation occurs if concrete is dropped from a height. This is usually restricted to not more than six feet unless chutes are used, in which case concrete can be placed from a much greater height.

All structural concrete is required to be vibrated to produce a dense homogeneous mass. Immersion vibrators of the poker type are normally used and these must not be allowed to come into contact with the reinforcement.

The location of construction joints are either shown on the drawings or may be as agreed by the engineer. This can be very important in relation to the design of the structure. The treatment of the surface of the concrete before placing new concrete against it is fully described in the specification.

Reinforcement must be placed and maintained in the position shown on the drawings. It is particularly important that reinforcement is not displaced to such an extent that the clearance between the bar and the formwork is reduced. This would result in the cover of concrete over the steel being reduced, with the possibility of corrosion of the steel taking place and ultimate spalling of the concrete.

The specification requires that, unless a particular type of surface finish is shown on the drawings or specified elsewhere in the contract documents, the permanently visible surface of the concrete is to have a smooth and dense finish. Minor blemishes are to be removed by rubbing down the surface immediately after striking the formwork.

Pre-stressed concrete. Special requirements are normally quoted in the specification depending on the design of the structure and the method of stressing concerned. A higher strength of concrete is required for this type of construction. For pre-tensioned units a minimum strength of 7,500 lb. per square inch at 28 days is frequently specified and similar strength may also be required for certain types of post-tensioned work.

Riveted and welded steelwork. The specification for structural steelwork for bridges is largely based on B.S.153 'Steel Girder Bridges' with considerable amplification and amendment to meet the particular circumstances. Welding requirements are covered by the appropriate British Standard depending on the particular type of steel to be used. An important feature of welding procedure is the requirement that welders must pass qualifying tests before being permitted to work on the structure.

Protection against atmospheric corrosion. The basic requirements for the protection of fabricated steelwork are given in B.S.153, previously referred to, and B.S. Code of Practice 2008 'Protection of Iron and Steel Structures from Corrosion'. A number of amplifications and modifications of these British Standards are given in the specification.

The surface preparation of the steelwork is required to be carried out by a specified blast cleaning process. After removing all loose or embedded grit or shot, and within four hours of the completion of the blast cleaning, a pre-treatment etch primer is to be applied by brush or airless spray. Two lead-based priming coats are then applied to the surface in the fabrication shops, and their combined thickness is to be not less than 0·003 inch. With the addition of an undercoat and finishing coat of micaceous iron ore paint, the combined total thickness is to be not less than 0·005 inch. If an additional undercoat and finishing coat are to be applied, which may be required in the case of the outer faces of external girders, the total paint thickness is increased to a minimum of 0·007 inch.

Materials and Testing

The specification contains separate series of clauses dealing with Materials and the Testing of Materials and Workmanship. Many of the materials are required to comply with British Standards, which usually detail the testing procedures to be adopted. Specific requirements may, however, be given for materials applicable to the particular contract and these include the aggregate characteristics for pavement construction. In addition to the Polished Stone Value for aggregates exposed in the running surface of carriageways, already mentioned, other requirements which are frequently specified are the Flakiness Index, Aggregate Crushing Value and Absorption Factor. The tests for these values are carried out in accordance with B.S.812.

Tests are also specified for items such as pre-tensioned pre-stressed beams. Loads in equal increments are applied to selected beams when supported at their design points of bearing. The deflections of the beams are measured and plotted against the loads. The load/deflection curve for each beam should show no appreciable variation from a straight line.

Limitations of space have only permitted an abbreviated look at the specification for a motorway contract. It is hoped, however, that an insight has been given into the large number of items of work which have to be covered and carefully described. Without it the designer's requirements cannot be fully understood, priced and executed by the contractor, and the resident engineer, with the responsibility for the supervision of the work, has no yardstick by which to judge its acceptability.

Bill of Quantities

The Bill of Quantities gives the quantities and brief descriptions of work to be performed. The quantities are computed from the drawings and show the extent of the work. The description identifies the item of work by reference to the appropriate drawing or clause of the specification. The bill is intended in the first instance to give information upon which tenders can be obtained. It must, therefore, cover every item of work involved, and all the obligations of the contractor.

In preparing a tender a contractor prices each item in the bill and arrives at a total value for the quantities included. This is the tender sum which is compared with those submitted by other contractors. It is usual in the case of motorway contracts involving the expenditure of public money for the lowest tender to be accepted.

When the contract has been entered into, the rates in the priced Bill of Quantities are applied in assessing the value of the work as carried out. For this purpose measurements are agreed between the resident engineer and the contractor's agent. Such measurements form the basis of both the interim and final payments. The method of measurement must be clearly understood by both parties. It is, therefore, stated in the I.C.E. Conditions of Contract that, unless the Bill of Quantities expressly shows to the contrary, it must be prepared in accordance with the standard procedure laid down in the Standard Method of Measurement of Civil Engineering Quantities.

Motorway contracts often extend over a period of several years. At the present time, when there are large and rapid variations in the cost of labour and materials, it is extremely difficult for contractors preparing a tender to forecast the cost of items of work so far in advance. It has therefore been the practice in recent years to include a series of 'Variation of Price Clauses' in the conditions of contract for contracts extending over a period in excess of two years. In the case of labour, the basis of variation is related to awards made by the recognised wage-fixing bodies of the trades concerned. For materials, the contractor is required to submit with his tender the basic prices of certain materials specified by the employer. Variations of the contract price are only made relative to the basic prices and on the quantities supplied to the contract.

Future trends in contract procedure

In June 1965 the Economic Development Committee for Civil Engineering formed a working party to examine costs and productivity in road construction. Their findings were contained in two reports, entitled *Efficiency in Road Construction*, published in the summer of 1966[1] and the spring of 1967[2] respectively, and already mentioned in Chapter 4.

It was considered that a number of defined aims should underly the future development of the road construction industry. Several of these aims related directly to the letting of contracts. These called for a greater continuity of work for plant and contractors by distributing it on a more rational basis to fewer firms and for more specialisation in major highway works.

The working party reached certain conclusions consistent with these aims and made recommendations which would help to achieve them. Based on the belief that not many firms can play an important role in the execution of major contracts, it was recommended that the contractors themselves should aim to concentrate the work in the hands of fewer teams. It was stressed that the Ministry should ensure that its selective tendering procedure enabled all firms to develop in standing and calibre, with unsatisfactory firms relegated or discarded. In this connection an irresponsible attitude to the submission of claims was a factor to be taken into account. It was considered that, in general, no more than about four tenders were necessary.

It was further recommended that the Ministry should consider the possibility of contractual arrangements which would give incentives for early completion as well as a liability to damages for late completion.

It was suggested that better information might be given to contractors about ground conditions and about the sources of natural materials known to be acceptable.

The means by which the practical experience of contractors could be utilised also received consideration. It was felt that, at the tender stage, contractors should not be discouraged from submitting alternative tenders. In addition, engineers should be encouraged to discuss suggestions with a contractor who believes he can offer a significant saving in cost and, subject to suitable safeguards, to adopt them.

[1] National Economic Development Office, *Efficiency in Road Construction*, London, Her Majesty's Stationery Office, 1966.

[2] National Economic Development Office, *Second Report on Efficiency in Road Construction*, London, Her Majesty's Stationery Office, 1967.

It was recommended that the system of assessing and certifying interim payments should be reviewed in order to reduce the call on the manpower involved.

The Ministry was urged to carry out an experiment in serial contracting. This would require a study to determine the optimum size of projects and their phasing.

In the second report, one of the conclusions drawn was to the effect that there were no advantages in adopting any other form of contract from that generally in use. Any extension of what is known as functional contracting was not advocated. This arrangement involves letting individual contracts for different parts of the work. For example, bridgeworks, earthworks and paving would be carried out as separate contracts, with the engineer responsible for the overall management.

The extent to which it will be possible to put these recommendations into effect will remain to be seen.

Chapter 10

The construction of the motorway

Some aspects of construction

In order to give some indication of the sequence of operations and a few of the problems involved in the construction of a motorway, the following describes briefly the various stages of the work:

Alterations to mains and services

The construction of a length of motorway inevitably interferes with the existing mains and services of statutory undertakers, including electricity and telephone cables, gas and water mains, and sewers. The extent of the alterations involved can vary considerably and obviously the problem is greater in urban areas than in the case of a rural motorway.

Since they have the responsibility for maintaining services to their consumers during construction, the statutory undertakers are naturally reluctant to permit outside agencies to interfere with their apparatus, which in many instances is of a specialised nature. They, therefore, either carry out the work themselves or appoint specialist contractors.

The time taken to deal with the alterations may be considerable and is often difficult to estimate. The effect on the motorway contractor can be serious if delay is caused and it is vital that in the preparation of the scheme the engineers should investigate fully all the aspects of the work involved, in consultation with the undertakers concerned.

As far as possible the work should be carried out in advance of the construction of the motorway. It quite often happens, however, that the diversion of a main or service cannot be undertaken until part of the motorway construction work has been completed. For example, the construction of an overbridge carrying a road may also be required to accommodate a diverted service. The permanent alteration cannot, therefore, be completed until the bridge has been built but, in a case such as this, the temporary diversion could be carried out well in advance.

Although a considerable amount of pre-planning may have been carried out before the motorway contractor has been appointed, it is vital that all these operations are co-ordinated with the contractor's programme. In many instances the timing of the statutory undertakers' work may, in fact, be the controlling factor in the formulation of the programme. It is vital that tenderers are provided with as much information as possible in this connection.

The early work

It has been the practice in the past to allow in the Bill of Quantities for the provision of a temporary fence to enclose completely the acquired land. This is normally of a type which can be erected rapidly as soon as the initial setting-out has been completed.

The contractor is responsible for the provision of all working areas which he requires outside the limits of the acquired land. Many contractors have endeavoured to obtain from the adjoining landowners the temporary use of a strip of land 30 feet wide or thereabouts on each side of the motorway and, where this has been possible, the temporary fences have been erected enclosing the strip.

The first major operation involves stripping the top soil and this is normally carried out by scrapers. The soil is either stockpiled on the working strip or in other areas specially acquired by the contractor for the purpose. After allowing for the final soiling of slopes, verges, etc., there is generally a surplus and the contractor may be required to offer this for sale, in the first instance, to local farmers.

In order to facilitate the carrying out of the earthworks, it is generally advisable to carry out as much of that part of the drainage work as possible, which will prevent water entering the site. An early operation should be the laying of the cut-off drains at the top of cutting slopes to intercept field drains which will be severed by the excavation. In addition, it may be desirable to lay the French drains to intercept surface run-off where the land falls towards the motorway and this should be done in the case of both cuttings and embankments.

Equally urgent is the construction of culverts across the line of motorway to take the flow of ditches and other watercourses. These may vary in size from a single pipe to a multi-plate arch built up from corrugated steel plates. The early completion of this work is vital to ensure that as far as possible there should be no interruption to the movement of constructional plant throughout the site.

Bridges

The speed at which bridgeworks can be carried out may be the controlling factor in the successful completion of a motorway project. Bridges are often necessary in order to provide access to the various sections of the site and, although temporary structures may be utilised, they add to the cost of the work.

Earthworks

The start of the earthmoving operations will depend to a large extent on the time of the year. It is generally recognised in Britain that, where clays or similar materials are involved, the earthmoving season will start in about March and may extend to the end of October. Outside these months the rate of evaporation is so slow that if the site becomes wet, which is almost inevitable in our climate, it never really dries out until the spring. There is no reason, however, why free draining materials such as rock and gravels, which are less affected by the weather, should not be excavated and placed in embankments throughout the winter months.

There are, of course, exceptions. During the severe winter of 1962–3, in the construction of the Preston-Lancaster section of M.6, a considerable quantity of unsuitable clay was excavated and moved to tips. The hard frost enabled equipment to move easily across country.

The type of plant used in earthmoving is dependent on two major factors–firstly the type of material and secondly the length of haul involved. If it is necessary to use public roads, lorries must be used, in which case excavators will be employed for digging and loading. On short hauls within or adjacent to the site, tractors and scrapers may be employed as their slow speed of travel is not too great a disadvantage. However, over the longer distances, the much faster rubber-tyred scrapers are more efficient.

The ground conditions may be such that both types of scraper are unable to operate without becoming 'bogged-down'. It may, therefore, be necessary to use drag-line excavators loading into lorries.

The difficulty with all these methods is the effect of wet weather. The surface of the comparatively large area over which the excavation is taking place is quickly affected and work will usually be held up until the surface dries out.

An alternative, in the case of deep excavations, is to employ a face shovel. The material in the almost vertical face is less affected by rain and any interruption in progress is very much reduced.

The critical factor, however, is the condition of the embankment or of the tip where the material is to be placed. The importance of maintaining a shape which will allow the rapid run-off for rainwater cannot be over-emphasised.

Drainage

All the drainage works should be completed before the protective layer is removed and the formation is finally trimmed. Otherwise the formation will be affected by the weather and by the movement of construction traffic. The excavation width of the trenches should, as far as possible, be kept under strict control. An excessive width will affect the stability of the surrounding ground and, in particular, the formation under the carriageways. In addition, trenches excavated into embankments may create points of weakness which can possibly lead to slips. The backfilling of the trenches should be carried out at the earliest opportunity. Apart from the possibility of deterioration of the trench, the excavated material will weather and may be rendered unsuitable for backfilling.

Carriageway works

With the completion of earthworks and drainage, work can then proceed on laying the sub-base in the case of flexible construction, or the base where rigid construction is to be employed.

It is an advantage if the thickness of the first layer is sufficient to ensure that, when placed, it is capable of carrying the lorries bringing the material for the next layer. However, it must not be too great to prevent it being adequately compacted. The materials used for sub-base are often prone to segregation in loading, transporting and depositing. One method of dealing with this problem is to tip the load on to the top of the layer already placed and push it out ahead. This has the effect of ensuring that the larger particles are rolled forward and lie on the formation. The smaller particles then fill the interstices from the top.

Where concrete marginal strips and drainage channels are to be provided, these are normally laid before the sub-base is finally shaped and trimmed. Various methods of forming the marginal strips and channels have been employed, from the use of conventional forms to machines capable of extruding the concrete. Once these have been completed, they form a ready datum from which to lay the following pavement courses.

Self-propelled paving machines are used for laying the base and surfacing materials in flexible and semi-flexible construction. The thickness of each layer depends on the type of material and the size of aggregate.

The future trend is for this type of equipment to operate under the control of wire-guided electronic mechanisms. Using this method, concrete marginal strips are no longer required for fixing the line and level of pavement courses. Instead a tightly stretched wire held above the surface at either side of the machine run is accurately set to line and level. Feelers attached to the machine and in contact with the wires provide the means of control.

This principle has also been adopted in the operation of the 'slip form paver' for laying concrete pavements. Concrete is fed into the front of the machine which moves continuously forward, and it leaves behind the concrete carriageway without the necessity for providing side forms or bankettes.

Formation trimmers have also been developed capable of producing a high degree of accuracy. The use of this equipment means that the formation level can be trimmed to such a level that no more than the specified thickness of pavement need be laid, with a resultant saving in cost.

Completion

There are many other operations to be carried out before the motorway can be opened to traffic–the erection of the permanent fencing and safety fences where appropriate; the provision of white lines and reflecting road studs; the erection of traffic signs and their lighting; laying the cables for the emergency telephones and erecting the telephone cabinets, to name but a few.

Examples of completed motorways

At this stage I should like to refer to the construction of some of the motorways in Lancashire, for which I have been responsible to the Ministry of Transport. (See Figure 31.)

Preston By-pass, M.6

Construction work on the Preston By-pass, Britain's first motorway, started in June 1956. This comparatively short length of $8\frac{1}{4}$ miles was designed to serve initially as a by-pass of Preston and its neighbouring industrial and urban areas. Its ultimate function, however, was to form part of the North-South Motorway, M.6, and it was the forerunner of several motorway projects in Lancashire which led to the completion of the $61\frac{1}{2}$ miles of the M.6 within the County in January 1965.

Preston By-pass was due to be completed within two years, but because of bad weather the contract period was extended by five months and the road was opened to traffic in December 1958 by the then Prime Minister, Mr. Harold Macmillan.

In comparison with present-day undertakings it might be considered small, but as a civil engineering project it was a scheme of considerable magnitude.

The topography of the area through which the road passes was generally undulating in character. In crossing the valleys of the Rivers Ribble and Darwen, deep cuttings

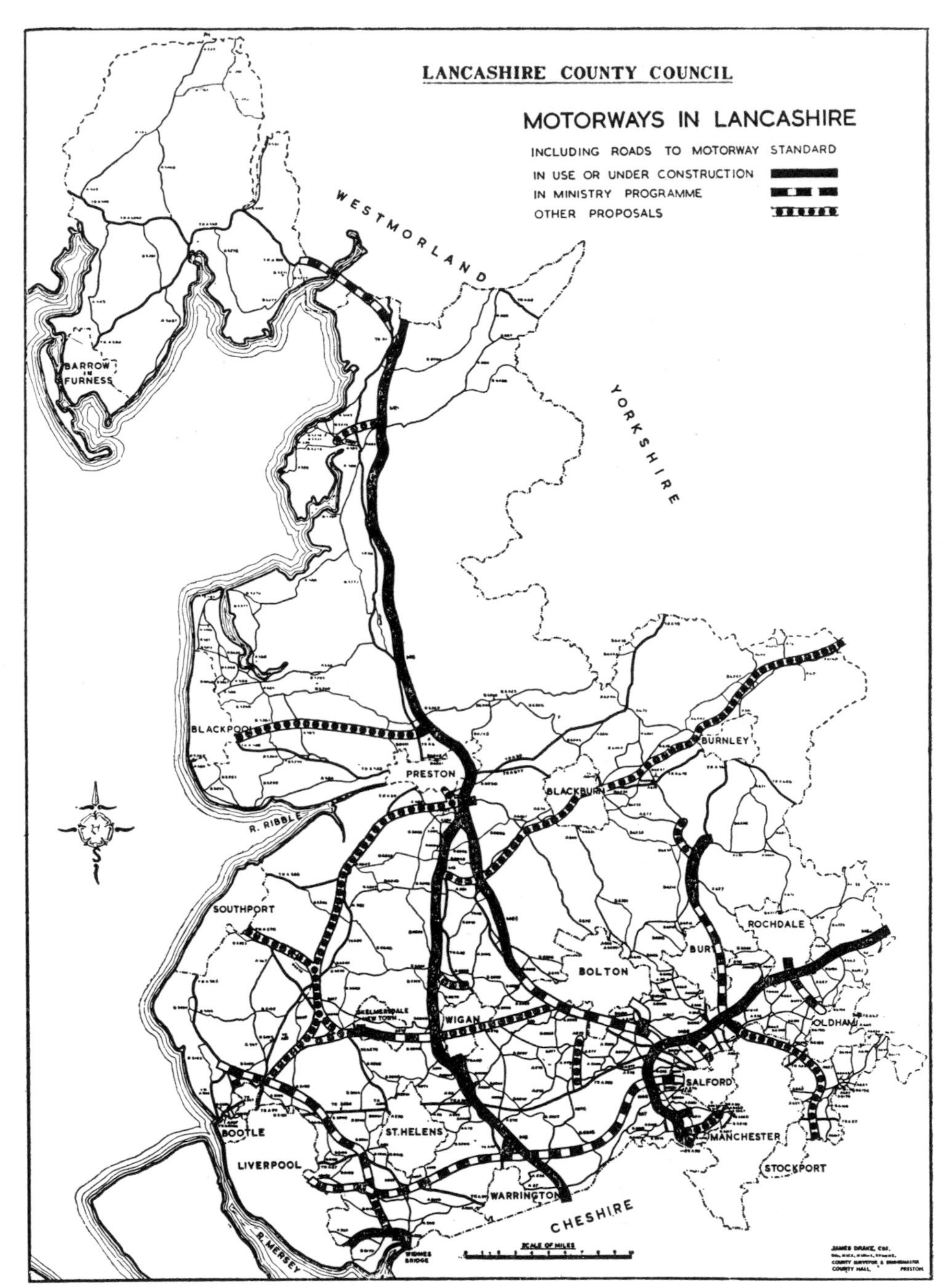

Fig. 31. Motorways in Lancashire

and high embankments had to be formed in order to adhere to the limiting gradient of 1 in 25. Over three million cubic yards of material had to be excavated, a substantial part of which was concentrated on the edge of these valleys where there were high escarpments. The most severe was the north escarpment to the Ribble valley, where a cutting 60 feet deep had to be excavated, the material obtained being used to form an adjoining embankment 60 feet high.

An important feature was the very bad ground conditions encountered in the earthmoving operations. A soil survey at the design stage had indicated that the soil types would be predominantly clays with wet sand and silt veins. This information was found to be substantially correct, but the wet sand and silt veins were very numerous and randomly dispersed. The effect of these and the continued wet weather during the earthmoving period led to atrocious conditions probably unequalled elsewhere on any other motorway contract. The result was that large quantities of excavated material were rendered unsuitable and to replace the deficiencies a considerable amount of imported filling had to be brought to the site.

Because of the nature of the material, great difficulty was experienced in establishing a satisfactory formation, particularly at the northern end of the By-pass. By punching large rock into the formation a stable condition was produced. This technique, known as stabilisation, has been adopted generally for dealing with formations of this type.

The sub-base for the carriageways consisted of a variable depth of burnt red shale obtained from colliery waste heaps throughout Lancashire. A nine-inch thick 'wet-mix' base was laid. This consisted of crushed and graded limestone mixed with a controlled percentage of water to assist in binding the material together and to facilitate mechanical laying. The surfacing was $2\frac{1}{2}$ inches of tarmacadam base course overlaid with $\frac{3}{4}$-inch of fine cold asphalt as a wearing course, into which pre-coated chippings were rolled.

This surfacing was only temporary, and it was intended that a permanent surface would be laid when all settlement had ceased. Six weeks after opening to traffic a small amount of frost damage occurred as a result of an exceptionally quick thaw, after a prolonged period of very severe frost, when there was a rise in temperature from 8°F., to 43°F. within a period of 36 hours. Following the publicity given to this incident, it was decided that the four-inch thick final hot rolled asphalt should be laid at an earlier date than originally intended and this was completed by the end of 1959.

From that date the principle of laying temporary surfaces on motorways was abandoned, and it is now the practice to lay the permanent surface before opening to traffic.

The original Ministry of Transport specification for the hard shoulders on Preston By-pass consisted of 4½ inches of gravel, on which a 4¾-inch thickness of a mixture of stone, sand and loam was laid. The surface was then sown with grass seed. In the latest type of construction the sub-base is carried across from the carriageway and under the full width of the hard shoulder. The depth of construction is, therefore, the same as that of the carriageway, but some of the materials used are not quite to the same standard.

In 1963 the hard shoulders on Preston By-pass were reconstructed following these principles. An upstand was formed along the outer edge to form a channel for surface water, which flows into a piped drainage system via gullies. The surface is dressed with red schlamme.

Only a very short period was available for this work. As valuable time would be taken in inviting tenders and because of traffic considerations, the work was undertaken by the Direct Labour organisation of the Lancashire County Council and carried out at a rate of over a mile per week, the 12 miles of hard shoulder involved being completed in under three months.

Although the original layout of the By-pass provided a sufficient width in the central reserve to allow dual three-lane carriageways to be built, only dual two-lanes were incorporated in the first instance.

The construction of the third lanes was carried out by contract during the latter part of 1965 and completed early in 1966.

All the bridges on the By-pass were individually designed to meet the engineering requirements. A total of 22 in number was required and great care was taken to ensure that each type was aesthetically suitable for the particular site.

Two major bridges were required. The Samlesbury bridge, which carries the By-pass over the River Ribble and Trunk Road A.59, is shown in Plate 8. The Higher Walton bridge crosses the valley of the River Darwen and, with a total length of 474 feet, is the longest bridge on the By-pass. Of a simple viaduct type construction, it has continuous steel plate girders carrying a reinforced concrete deck over six spans.

Thelwall to Preston, M.6

In September 1959 work on this 27-mile length of motorway started with the construction of the Thelwall bridge and approaches.

The main bridge is described in Chapter 8, but the contract also included the construction of part of a double U-type interchange, a new two-span bridge carrying A.57 over the motorway and approach embankments and roadworks.

The embankment carrying the motorway and slip roads at the northern end of

Thelwall bridge was built over an area of soft alluvial deposits up to 30 feet deep. This layer of mainly sand and silt mixed with clay was first stabilised by means of 15-inch diameter vertical sand drains driven at ten feet centres over the whole area covered by the embankment. The total length of the 3,300 drains involved was 77,000 feet.

A two-feet thick layer of pervious material was then laid over the area prior to forming the embankment. Filling was carried out at a controlled rate in order to avoid setting up excessive pore water pressures in the underlying material. These pressures were measured by means of instruments known as 'piezometers' connected to pore pressure cells placed at various points under the embankment. Settlement gauges were also installed at the site of each cell.

The effect of the loading imposed by the embankment was to force water out of the alluvial deposit into the sand drains and thence up to the horizontal pervious layer where it was able to discharge.

On completion of the embankment, which varied in height from 15 to 35 feet, a surcharge of ten feet of filling was added to provide additional loading. This was left in position until the rate of settlement became negligible, and it was then removed.

Approximately half a million cubic yards of imported filling were necessary to form the embankments on the two approaches, the higher being at the south end of the bridge where it has a height of 55 feet.

The carriageways on the approaches, slip roads and A.57 diversion are of flexible construction. On the section of motorway and on the slip roads the base consists of a ten-inch thickness of bitumen bound granular base, laid on a crusher-run limestone sub-base. On the A.57 diversion a similar sub-base is overlaid with a nine-inch thick base, comprising six inches of 'wet-mix' superimposed by three inches of bitumen bound granular base. A temporary surface of $2\frac{1}{2}$ inches of close-textured bitumen macadam and surface dressed was provided on A.57, while the final four-inch thick hot rolled asphalt surface was laid on the section of motorway and slip roads.

Another contract, started at the same time, involved the construction of the Gathurst viaduct carrying the M.6 and located approximately three miles west of Wigan. This bridge is also described in Chapter 8.

Work on the main length of motorway began in February 1961, when contracts were let for the remaining 26 miles.

As indicated in Figure 32, the motorway has an overall width of 124 feet, accommodating dual three-lane carriageways, hard shoulders, each ten feet wide, plus an 18-inch wide channel at the back, a central reserve 15 feet wide including the 12-inch marginal strips, and side verges six feet wide. A 12-inch white painted line separates the carriageways and hard shoulder and, of this, four inches are reflective. On the central reserve side of the carriageways a four-inch wide reflective white line was also provided, additional to the 12-inch wide concrete marginal strip.

A pleasing free-flowing alignment has been obtained with a minimum radius of curvature of 3,580 feet and a gradient no steeper than 1 in 33.

The soils on the line of the motorway consist of clays of varying plasticity, soft and hard grey shales, dry and very wet sands, silt, peat and sandstone–and coal!

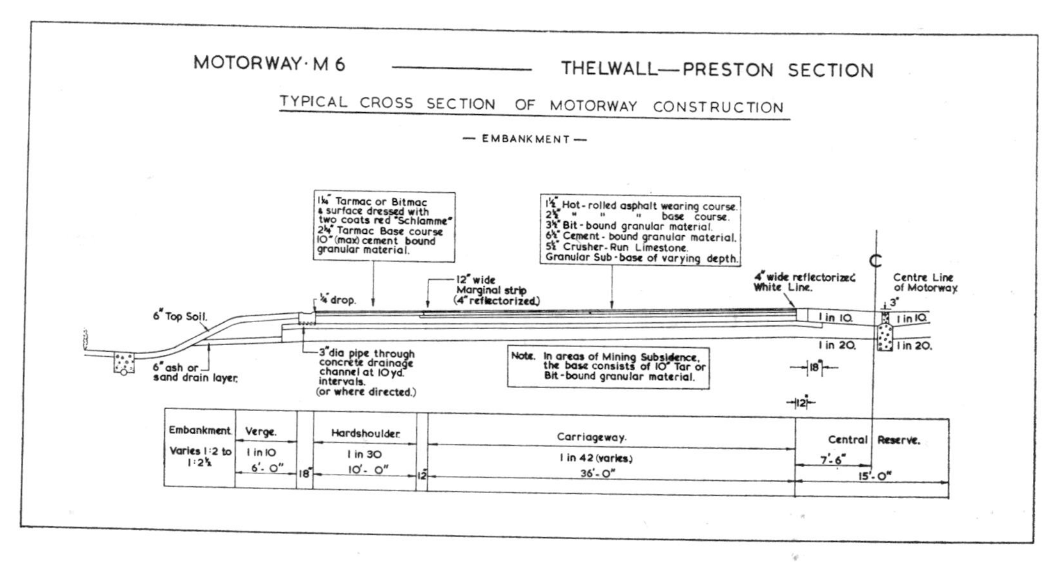

Fig. 32. Typical cross-section for Thelwall to Preston length of M.6

From near Warrington to Charnock Richard, a distance of some 19 miles, the motorway passes over coalfields and, in a number of places, the seams were exposed during excavation. In fact a total of 13,000 tons of good quality coal was excavated and handed over to the National Coal Board for sale.

Shallow, old and unrecorded mine-workings were met on several occasions (Plate 17). Once located, they were either dug out and back-filled if within 15 feet of the road or, where they were between 15 and 30 feet deep, a reinforced concrete protection slab was provided. Old mine shafts were also encountered, 25 being found within the limits of the motorway, varying in depth from 60 to 900 feet (Plate 18). Where the old shafts had not been completely back-filled, selected material was deposited in them and in one case it was necessary to provide additional filling to a depth of nearly 160 feet. Shafts lying under or near to the carriageways were covered with reinforced concrete rafts.

Subsidence from current and future mine-workings is expected over a length of 9½ miles and the carriageways, drainage and eight bridges have been designed to cater for settlement of up to 13 feet by the 1980's at the worst point.

Approximately eleven million tons of material were excavated and more than three million tons of fill were imported to make up the deficiency of suitable material required to form the embankments.

Surface water from the motorway carriageways and hard shoulders is collected into concrete channels and discharged by gullies into the main piped surface water drainage system in the hard shoulder or central reserve. In cuttings French drains are provided in the side verges, central reserve and at the top of the slopes where the slope of the natural ground is towards the motorway. On embankments a French drain is provided in the central reserve and at the bottom of the embankment slopes.

Where mining subsidence is anticipated, heavy duty reinforced concrete pipes with flexible joints were used. Culverts in these areas and in many other instances were of corrugated steel sections.

The carriageways are of flexible construction designed to carry the heavy loading specified by the Ministry of Transport and the overall construction depth varies from 12 inches in rock areas to 41 inches according to design requirements. The sub-base consists of granular material overlaid by a seven-inch layer of crusher run stone.

A self-propelled trimming machine, known as a 'Flicker', was used to strike off the surface of the crusher run sub-base to the required tolerance, by means of rotating paddles. This machine ran on the *in situ* concrete marginal strips and drainage channels, which had previously been formed to correct line and level.

The majority of the base consists of 6½ inches of cement-bound granular base overlaid with 3½ inches of dense bitumen bound macadam. On sections liable to be affected by mining subsidence the base is ten inches of dense bitumen bound macadam. The surface of the carriageways consists of two courses of hot rolled asphalt, comprising a base course 2½ inches thick and a 1½-inch thick wearing course with precoated granite chippings applied to the surface.

Apart from the two viaducts at Thelwall and Gathurst, described in Chapter 8, there are 79 further bridge crossings, 75 on the motorway proper and the others on the link roads. Several of the bridges are illustrated.

This length of motorway was opened in July 1963 by Mr. Ernest Marples, the then Minister of Transport.

The Preston-Lancaster Motorway, M.6

The 13½ miles of motorway between the Preston and Lancaster By-passes was the fourth and final length of the M.6 to be constructed in Lancashire. It has no intermediate interchanges and is one of the longest lengths of motorway between interchanges in Britain.

The connection to the Preston By-pass at its southern end, the Broughton interchange, has already been described in Chapter 7. At the northern end at Hampson Green, where the motorway joins the Lancaster By-pass, a more conventional 'trumpet' type junction has been built.

The overall effective width of this length of motorway is 129 feet, made up of dual 36 feet wide carriageways, a 15 feet wide central reserve including marginal strips, ten feet wide paved hard shoulders and ten feet wide side verges. The minimum radius of curvature is 3,820 feet and the maximum gradient 1 in 47.

Alternative tenders were invited for three different types of carriageway construction–flexible, semi-flexible and rigid. The lowest tender, involving the construction of a rigid type pavement for the carriageways on approximately 12 miles of the length and semi-flexible construction at the interchanges and on the approaches to underbridges, was accepted (Figure 33).

The construction of concrete carriageways on major projects of this kind requires the use of several items of mechanical equipment, each carrying out a different operation in succession. When assembled together in line they are generally referred to as a concreting train. In order that the train should be able to operate as efficiently as possible, it is important that it should be able to move forward at a steady rate. It is, therefore, necessary for all the preceding operations to be completed in advance of the train and this principle applies particularly in the case of bridge construction both under and over the motorway.

Many new features were incorporated in the design of the concrete carriageways and the contractor adopted a number of methods unique to this type of construction in Britain.

In order to provide bankettes on which the concreting train could travel, the marginal strips were laid well in advance. The base of crusher run limestone was then trimmed with the same machine used for the sub-base on the Thelwall-Preston contract.

The reinforced concrete slab is 10½ inches thick, laid in two courses. The bottom 7½-inch thick course is of 1½-inch maximum gauge limestone concrete and the top course three inches of air-entrained concrete with ¾-inch maximum gauge granite aggregate. After trials, a mix with an aggregate/cement ratio of 5·7 : 1 was found to be suitable for the bottom course and a ratio of 4·8 : 1 was used for the top course. The water/cement ratio was restricted to a maximum of 0·50 : 1 in the case of both mixes.

The fabric reinforcement laid between the two courses has a weight of 8·95 lb. per square yard.

In order to give a high resistance to skidding, a deep surface texture was produced by transverse brushing the surface of the freshly-laid concrete with a steel wire brush, in addition to using granite aggregate with good skid resistance properties in the top course.

The spacing of the expansion joints, generally 480 feet, was reduced to 360 feet during the cooler autumn period of laying. The spacing of the contraction joints is 40 feet. Two longitudinal joints are provided in the 36-feet wide slab at the third

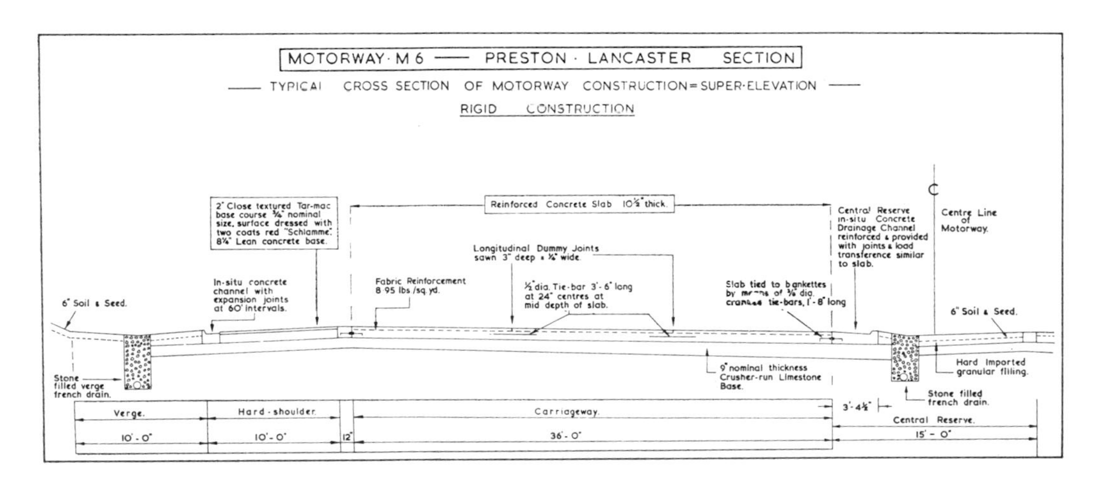

Fig. 33. Typical cross-section for the Preston–Lancaster Motorway, M.6

points. Dowel bars of 1¼-inch diameter mild steel are provided at 12-inch centres across the contraction and expansion joints, at the mid depth of the slab to provide load transference between adjacent slabs. Across the two longitudinal joints ½-inch diameter bars three feet six inches long and at two-feet centres, tie the slab together and these are also located in the mid depth of the slab. Between the marginal strips and the slab, ⅜-inch cranked tie bars at 15-inch centres prevent the joint opening. These were supplied in two halves threaded at each end and connected by means of a ferrule to enable one half to be cast into the marginal strips in advance of laying the pavement.

The grooves for the longitudinal, contraction and expansion joints were all cut by diamond-faced saws. For the longitudinal and contraction joints these are ¼-inch wide by three inches and 2¾-inches deep respectively. In the case of the expansion joints the groove is 1¼-inches wide by one inch deep down to the top of the parana pine filler board provided in these joints.

As already mentioned in Chapter 6, it is essential that the grooves for the joints are sawn early. Depending on the weather conditions, the contraction joints were sawn within five to 24 hours of the concrete being placed. Because timing of this operation has to be carefully judged so that the concrete is sufficiently hard to allow sawing without ravelling of the surface taking place, much of the sawing was carried out at night using floodlights.

On this section of M.6 the grooves were temporarily caulked with rope to prevent the ingress of grit. Subsequently this was removed, the groove was cleaned by sand blasting, dried and then filled with a rubber-bitumen sealing compound poured under pressure.

One of the main advantages of a concrete pavement is that it requires very little maintenance. At the same time it is necessary to ensure that the expansion and contraction joints, which are subject to temperature movement, remain satisfactorily sealed.

The concrete for the slab was produced in an electronically controlled central batching and mixing plant, comprising two 7½ cubic yard mixers, one for each type of concrete. Concrete was delivered to the train by 7½ cubic yard truck mixers which for this purpose were used as agitators.

The main feature of the concreting train was its ability to lay the full 36 feet width of slab in one operation. The maximum length of 36 feet wide carriageway laid in a 12-hour working day was 2,520 feet, a European record at that time.

Concreting was completed at the end of October 1964 thus enabling the motorway to be opened by the then Minister of Transport, Mr. Tom Fraser, at the end of January 1965, three months ahead of schedule.

The hard shoulders throughout the majority of the length were constructed with a base of eight inches of lean concrete and surfaced with a two-inch layer of ¾-inch nominal size close-textured tarmacadam dressed with two coats of red schlamme.

A total of 44 bridges was required on this contract, several of which are illustrated.

Chapter 11

Maintenance and operation of motorways

Maintenance

To keep a busy motorway safe and tidy is not as simple as one might imagine. Although the problem varies throughout the Country, it is possible to obtain a general idea of the work involved by considering the methods used by the Lancashire County Council, acting as agents for the Ministry of Transport for the maintenance of M.6 in Lancashire.

The length of M.6 in operation in Lancashire is 61 miles, together with 6 miles of slip roads, the total length of single carriageway being 134 miles. There are 159 bridges, of which the longest is the Thelwall Viaduct, 4,417 feet long. The highest embankment and the deepest cutting are 60 feet in height and depth respectively. The area of grass, including the central reserve, amounts to 807 acres. The mileage of drains to be kept free-running is 168, whilst 8,600 gullies have to be emptied regularly.

In charge of the maintenance is a Senior Engineer (Motorways) assisted by three Clerks of Works. The senior engineer also has other duties.

It is Ministry policy to build maintenance compounds at approximately 12-mile intervals. In the earliest motorways the compounds were sited alongside the motorway, generally at service areas, but later policy, with which I fully agree, is to site the compound off the motorway but near to a junction.

Each maintenance compound provides accommodation for the necessary maintenance vehicles, plant and signs; also offices for the staff, stores, fitters' shop and facilities for the routine maintenance of vehicles. In addition there is a mess room for the workmen and a bunk room, for stand-by duty in winter. There is also a hard-standing area where 1,500 tons or more of ground rock salt can be stored, and generally two 40-ton capacity hoppers for the quick loading of salt into lorries.

In each maintenance compound there are stocks of signs, danger lamps, traffic cones, fog flares and floodlighting equipment, and a trailer is kept fully loaded with this equipment ready for an emergency. The layout of a typical Ministry of Transport compound in Lancashire is shown in Plate 19.

Keeping the motorways functioning safely in frosty and snowy conditions is a top priority. Frost and snow warnings are given by the Meteorological Office, to supplement which the Clerks of Works make contact with the Meteorological Office at least twice a day between November and March.

During the depths of winter the men stand by at the compounds throughout the whole of the 24 hours. This is essential for a quick turn-out and also saves needless applications of salt. Patrols are undertaken at suitable intervals during the night since, although frost may not be general, isolated patches may occur and can be dealt with immediately.

The special vehicles used for gritting, salting and snow clearing have been developed under the direction of the Ministry of Transport's mechanical engineering section. The vehicles (see Plate 20), of which ten are in use on the Lancashire section of M.6, are six-wheel-drive machines fitted with special gritting bodies of 10 to 11 tons capacity. The vehicles are also fitted with snow ploughs and this operation, as well as salting, is hydraulically controlled. They are capable of applying salt over the full width of the 36 feet carriageways at speeds of up to 40 miles per hour, at rates of spread varying from half ounce to three ounces per square yard. The light application is usual for normal conditions and the heavier application for varying depths of snow.

It is essential that the slip roads and junctions receive the same treatment as the motorway proper and to ensure this, individual routing is laid down. In the cab of each vehicle is a plan showing the routing for that vehicle.

The hoppers enable loading and reloading to be carried out within seconds, and, applying the salt at approximately one ounce per square yard, the gritting vehicle is capable of covering 13½ miles of dual two-lane carriageway within 40 minutes of leaving the compound, or nine miles of dual three-lane carriageway within 27 minutes. About 2,500 tons of salt were used on the Lancashire length of M.6 during the winter of 1965–6, but in 1966–7 only 1,600 tons were used. When it has not been possible, by means of salting, to prevent snow from remaining on the carriageway, the snow ploughs on the front of gritting vehicles are used to clear it. To clear the carriageway of a two-lane motorway two vehicles are used, the one in the inner lane with its blade angled to push the snow to the nearside and the other in the outer lane with its blade angled in the other direction so as to deposit the snow on the central reservation. On a three-lane motorway the nearside and centre lanes are dealt with by two vehicles working in 'echelon', that is the vehicles move forward at the same speed with the one in the centre lane slightly ahead. Both vehicles have their blades angled to the nearside. The middle-lane vehicle pushes the snow to the nearside and the vehicle in this lane then clears all the snow and pushes it on to the hard shoulder, clear of the carriageway. The third lane is dealt with by pushing the snow on to the central reservation. Once the carriageway has been cleared work then proceeds on either clearing the hard shoulder completely or creating lay-bys to obviate the danger of vehicles stopping on the carriageway in an emergency. All the ploughs are fitted with rubber 'squeegee' attachments at ground level so as to avoid leaving a thin layer of compacted snow on the carriageway, or damaging the road studs.

For dealing with fog, over 600 flares of the kettle-type are available and the maintenance men's duty is to place them at junctions of all entrance and exit slip roads, and over any of the larger bridges which are known to be fog prone as soon as possible after fog warnings are received.

There are numerous other items of maintenance, all of which, together with the winter maintenance, have to be carried out within the sum allowed by the Ministry of approximately £2,300 per year for each mile of dual three-lane motorway and £2,100 for a dual two-lane motorway.

I believe it is in the national interest that the most important roads of the Country–the motorways–should be maintained in a tidy condition which encourages tidy driving, as well as creating a good impression with foreign visitors. We in Lancashire take great pains in this direction. Not only is the motorway swept regularly by mechanical sweepers, but also inspections are made, to remove dangerous debris.

The maintenance of grassed areas is an expensive problem during the spring and summer and is made more difficult due to the steep slopes to the cuttings and embankments. We have been experimenting over several years in developing machines specially suited for use on these steep slopes, which mulch the grass and do away with the necessity of raking and carting it away. A prototype of the latest of these machines is shown in Plate 21.

Another feature of British motorways which in these early years presents a problem, is the extensive tree-planting carried out by the Ministry of Transport. In all, more than 311,800 young trees have been planted on M.6 in Lancashire alone, and further areas are still to be planted. Unfortunately, it is not possible to cut the grass around the trees with the larger mowers and it has either to be cut by hand or by a small mower working on the hovercraft principle.

The large number of signs and the considerable length of guard rails also have to be kept clean. In Lancashire the former are washed at least four times a year, using for the larger signs, a Simon hoist with a hydraulic platform. This machine is fitted with special washing equipment devised by the engineer in charge of the County

Plant Workshops. It can spray clear water, or a detergent, or a combination of both. The hoist is also used on bridge inspection and maintenance.

The cleaning of the guard rails is carried out by a special washing attachment fixed to a sweeper-collector. This can deal with guard rails immediately adjacent to hard shoulders or carriageways, but where verges intervene, a gully emptier specially converted for the purpose is used. This machine also cleans the marker posts and the smaller signs.

As there are eight white lines, either full or dotted, on dual three-lane motorways (six lines on dual two-lane carriageways), their renewal is a fairly costly item. The lane markings are generally of a white reflectorised thermoplastic material, whilst the edge lines are formed with calcined flint chippings, or in Lancashire's case in paint, reflectorised with glass beads (ballotini). On the concrete carriageway between Preston and Lancaster we have used the same paint on the lane markings.

The renewal of painted lines can be very speedily carried out, using lorry-mounted spraying equipment. The lorry travels at between five and six miles per hour and the paint used is very quick-drying, which permits traffic to run over the new lines within 30 seconds of application. Until recently the renewal of thermoplastic lines has been a much slower operation, but a new machine has recently been produced which can spray the material through jets in the same way as the paint is applied and the speed of application is four to five miles per hour. Little maintenance is required to the 'cats'-eyes' reflective road studs.

An item of maintenance which, if neglected, will have a serious effect on the life of the carriageway, is keeping the drains free-running. All main drains and culverts are inspected and the 8,600 gullies are emptied every three months. Particular attention must be paid to the drainage before winter, because a wet sub-base might freeze in periods of frost and result in lifting of the surface. Heavy traffic using the road during a quick thaw could then damage the road.

One item of maintenance which Lancashire has been spared, and which will probably cease on other motorways, is the respraying of the hard shoulders. The Ministry's recent policy of substituting a slurry seal or other approved dressing instead of tarspraying the hard shoulders is welcomed. This type of dressing will require renewal at much less frequent intervals and, when required, can be carried out at a speed of $1\frac{1}{2}$ to 2 miles per day without having the danger of loose chippings.

All Lancashire bridges are inspected at six-monthly intervals; more frequently if required by special circumstances. A report is required for each inspection describing the condition of all the various parts of the bridge. As the motorway bridges are all of recent construction, they do not require any measurable amount of repair or remedial work, but a small expenditure is incurred annually on such items as the repair and adjustment of expansion joints, greasing of bearings and maintenance of drainage systems. Also the steel bridges require routine repainting, generally every five or six years, and a programme for this work is undertaken during the summer months. Effective maintenance can greatly prolong the life of bridges.

However well a motorway has been constructed there will come a time when resurfacing or major repairs will be required. As long lengths of motorway have been built at the same time, it is more than likely that the repair works will extend over several miles. Traffic conditions throughout the year will be taken into account in deciding when the work should be done and this may be restricted to week days or weekend according to circumstances. The work must be carefully planned, not only to minimise the risk of accidents to the workmen and the users of the motorway, but also to complete at one time as much work as possible in the carriageway cordoned off. As a last resort, one of the carriageways may have to be closed, but when the third lane was added to the Preston By-pass and Lancaster By-pass this was carried out quite expeditiously and without closing a carriageway, by allowing the contractor the use of the offside lanes of both the original two-lane carriageways over a maximum distance of 1½ miles at a time. The design of these by-passes had provided a wide central reserve to allow for the ultimate widening to dual three-lanes.

Policing and emergency services

Emergency services and policing on the motorways follow a general pattern and involve the Police, Fire Brigade and Ambulance. The provision and operation of these services requires the closest co-operation between the staffs of the Chief Constable, the Chief Fire Officer, the County Medical Officer of Health and the County Surveyor. This spirit of co-operation has existed on motorways in Lancashire right from the start.

To enable help to be summoned quickly in an emergency, telephone cabinets are provided in pairs opposite each other on M.6, at one-mile intervals. The telephones are sited in the verge at the back of the hard shoulder and the cabinets are identified by a number and a letter. The number signifies the distance from London in miles, for those motorways radiating from London. On the M.6, the letter A is on all the cabinets on the north-bound carriageway and B on the cabinets on the south-bound carriageway. This system gives quick and easy reference to the location of an incident.

Calls from motorway emergency telephones are connected direct to the Information Room at the Lancashire County Police Headquarters. If the fire or ambulance services are required, the details are relayed to their respective headquarters. In the case of breakdowns the Police Information Room informs either the breakdown service at the nearest service area, or nearby garages which undertake this work.

Police

As the Preston By-pass length of M.6 was the first motorway in Britain to be opened to traffic, the Lancashire Constabulary was the first police force to be faced with the problem of maintaining the smooth and safe flow of motorway traffic. Believing that the patrol and supervision of traffic on the Lancashire motorway required a separate traffic department organisation under a unified command, with specialist vehicles and highly trained personnel, the Lancashire County Police set up one and it has now been in operation for some years.

The procedure in dealing with an accident can be divided into seven stages, which the police have coded as 'ACE CARD': A–Approach from the rear, C–Caution signs, E–Examine scene; C–Casualties, A–Ambulance, etc., R–Remove obstructions, D–Detailed investigation. In the description of the procedure which follows, the two-man crew of a patrol vehicle–the driver and observer–operate as a team, with the driver in charge of the action taken.

A–Approach from the rear. In approaching from the rear, the patrol vehicle manoeuvre is started at least 900 yards away from the accident. Great care is needed, especially if the accident is on one carriageway and the police car is travelling on the other, as this involves crossing the central reserve at the nearest emergency crossing. On all emergency calls the flashing roof light of the vehicle is operated continuously on arrival at the scene.

C–Caution signs. Three sets of pairs of identical signs to warn approaching motorists are placed on both the hard shoulder at the nearside and the central reserve to the offside. The first, 'Police Accident', is placed about 900 yards from the scene of the accident; the second, 'Police Slow' is placed 600 yards from the scene and the third pair of signs 'Police Slow', 300 yards from the accident.

Cones, supplemented by orange and red warning lamps if at night, are placed in a diagonal line out from the edge of the carriageway, over a total distance of about 100 yards behind the police vehicle at the scene.

E–Examine scene. The driver assesses the situation, checks for any injuries and informs the police headquarters information room over the radio of any other services required.

C–Casualties. First-aid is given to any injured who are removed to the hard shoulder away from danger.

A–Ambulance, etc. The driver then checks with the information room that the ambulance or fire services have been called and, if required, makes a request for further aid. When the ambulance or fire tender arrives, it is directed by the police driver to a safe parking position and he meets the officer in charge of the vehicles to inform him about the incident.

R–Remove obstruction. The vehicles involved in the incident are removed to the hard shoulder at the earliest opportunity, where possible first marking the position with crayon so that necessary measurements can be taken later.

D–Detailed investigation. Statements are taken from witnesses if this can be done with safety. The property of any injured persons is safeguarded so that it can be returned.

This, then, is the procedure in the event of an accident on the motorway but, in addition, the police receive emergency calls from drivers whose vehicles have broken down. Breakdowns continue at a phenomenal rate. Notified calls to the information room in 1966 were 18,700 and in the peak holiday periods the number can reach 170 in 24 hours. The calls from motorists via the emergency telephones throw a heavy burden on the information room as experience has shown that an average of three to four outgoing calls are required for every one received. It is estimated that a total of over 65,000 telephone calls is made per year, dealing with motorway breakdowns and kindred matters. Only a very small proportion of these calls is concerned directly with police matters and the remainder are initiated by either negligence or ignorance on the part of the motorist.

A surprising number of vehicles are abandoned on the motorway, averaging in Lancashire approximately one per week on the 61 miles of the M.6. They are removed promptly, but storage facilities are limited and the vehicles are an embarrassment.

Considerable thought has been given to the selection of suitable vehicles for the police patrols. Up to 1967, the principal vehicle used was an estate car, supported by 650 c.c. motor cycles. Experience with multiple accidents in fog or ice conditions has made it necessary to use larger and improved warning signs of an obstruction ahead. This requires a great deal more equipment and a transit van has now been adopted. All the motorway vehicles are painted in fluorescent orange and fitted with a large roof 'Police' sign.

In the enforcement of the motorway regulations, the principal offences reported are those of stopping on the hard shoulder and speeding offences after the introduction of the 70 miles per hour limit.

The police patrols endeavour to enforce lane discipline by example, verbal warning and, as a last resort, prosecution. The worst offenders in this and other respects are the 'casual users' of the motorway, who appear to have little knowledge of the fundamentals of motorway driving or the motorway regulations.

The experienced motorway users, on the other hand, present few problems, whether they be drivers of goods vehicles, coaches or cars. They are conversant with motorway driving techniques and with the application of the regulations, and are aware of the limitations of their vehicles and of the dangers which can occur.

One of the biggest headaches to the police is that of warning approaching motorists that there is an obstruction on the carriageway, particularly in bad weather conditions. A large number of vehicles can be involved in nose-to-tail collisions under such conditions, resulting from drivers not adjusting their speeds to their range of visibility. To meet this hazard warning lights were introduced by the Ministry of Transport, which are provided at one-mile intervals to be switched on by the police in conditions of bad visibility, or when some other hazard exists. One of the first requirements at the scene of an accident at night or in fog is to provide floodlighting and the police vans now carry 1,000 watt projector lamps on 12 feet masts which are fed by petrol-electric generators carried in the vehicles.

Each incident in fog is dealt with as an emergency and the police have a routine arrangement with the Meteorological Office to keep them informed of the weather conditions twice a day. Whenever fog is forecast, a pre-arranged programme is put into effect whereby additional personnel report for duty at strategic points to operate road closures or diversions as necessary. Freezing conditions, whether associated with fog or not, require special measures. Ice warning signs of the drop-down flap type are permanently in position at slip roads and strategic points, and are displayed as soon as conditions occur which are favourable to the formation of frost. When either fog or ice is present, liaison is maintained with the service areas and adjoining police forces to give advance warning to motorists of the conditions they are likely to meet.

Fire

Although a Fire Authority's statutory obligation is limited to dealing with fires, the Fire Service has a tradition of being called to assist in all types of emergencies and most authorities consider they have a moral obligation to attend incidents on roads including motorways where, in any case, there may be a risk of fire. The fire service can respond quickly with a predetermined number of appliances. In addition to dealing with fires, their tasks include releasing persons trapped in crashed vehicles, removing spillage of inflammable liquids and helping to clear carriageways of wrecked vehicles, or loads from lorries or tankers discharged on to the road. These last sometimes consist of obnoxious or glutinous materials. The increasing number of such incidents, and the changes in the design of motor vehicles, which have made the rescue of trapped persons more difficult, have required the provision of special equipment.

An adequate water supply is, of course, essential for fire-fighting and a survey of water supplies over the whole length of motorway routes has been carried out so that maps can be provided to show all hydrants and static supplies having access points within 1,000 feet of each carriageway verge. To enable crews dealing with fires on motorways to locate quickly the nearest water supplies, indicator plates are being fitted so that they are visible from the motorway.

The introduction of motorways into the county area has produced special problems for the fire brigade and has necessitated considerable pre-planning to ensure that requests for the assistance of the fire brigade to any incident were made to fire stations which could provide the quickest possible attendance. In addition, it was also necessary to make arrangements with other fire authorities, whose administrative areas embraced sections of the motorway, to ensure that irrespective of administrative boundaries the nearest appliances were despatched to an incident.

A survey of the fires attended involving heavy goods and tanker vehicles indicates that the main causes of fire are the overheating of engines, axle bearings and tyres. It is therefore reasonable to conclude that high speeds over prolonged periods are the principal cause of overheating and indicate the need for vehicles using motorways to be mechanically sound and regularly serviced. Owing to the speed of traffic on motorways, accidents are frequently of the high impact type and this, in many instances, accounts for drivers and passengers being trapped in their cabs and cars.

Periods of limited visibility due to mist or fog provide additional hazards, including the possibility of 'concertina' crashes which, if not carefully guarded against, could occur when fire brigade appliances are actually deployed on rescue operations. To minimise this possibility, field experiments are proceeding and the searchlights carried on selected fire appliances have been modified by introducing a flashing unit into the electric circuit and fitting a translucent amber disc to the searchlights. On arrival at an incident, the searchlight is placed 100 yards to the rear of the fire appliance to warn oncoming traffic. This has so far proved a most effective warning.

The standard equipment normally carried on fire appliances has proved to be of value at many incidents encountered on motorways and on other roads, but in Lancashire it became evident that additional items such as hydraulic lifting and spreading equipment, cutting tools operating by compressed air and other special items were required, and the fire service has been specially equipped in this way.

Ambulance

The County Ambulance Service is responsible for providing coverage against accidents occurring along the whole of the M.6 in Lancashire, and covers motorway accidents from two separate control centres. There are ten ambulance stations which normally respond to motorway accidents and in addition, under the system of radio control, any other vehicles which may be in the vicinity can be diverted to the incident if the need arises.

Radio communication is well established in the ambulance service but recent technical developments have proved invaluable to ambulance crews dealing with serious accidents on the motorway and elsewhere. Where expert advice is required on the treatment of a seriously injured patient, the ambulance crew can be put into direct contact with the Casualty Officer at the Preston Royal Infirmary by means of a radio/telephone link. The latter, after being given a description of the patient's symptoms, can then give detailed medical advice to the ambulance crew as to the treatment of the patient during the journey to hospital. In the event of a major incident, hospital 'flying squads' are available and can be conveyed to the scene of the accident to give on-the-spot medical and surgical attention.

Aids to movement of traffic

Many aids to movement of traffic are provided on British motorways which assist the driver and contribute to safety.

Guard rail safety fences forming a continuous barrier are provided at various locations. About 2 feet 3 inches in height and usually formed of corrugated sheet steel, they are provided generally on both verges where the motorway is on embankment 20 feet or more in height and on the outer verge of embankments 10 feet high or more, where the radius of the curve is 2,800 feet or less. In addition, guard rails are provided at the approaches to all bridges and at any point of potential danger.

Various methods have been used to provide a safety barrier to prevent vehicles crossing the central reserve when out of control, including steel guard rails, turf mounds, steel wire ropes, dense shrubs and other means, but as yet no ideal method has been found. If a barrier is too rigid, there is a problem of rebound of a vehicle back on to the carriageway, possibly into the path of following vehicles.

All new motorway underbridges are provided with vehicle parapets which are designed of sufficient stiffness to act as a safety fence for vehicles. Previously, a lighter type of parapet was provided, supplemented by a steel guard rail. All motorway overbridges which carry vehicular traffic over the motorway (except occupation bridges for farms) are provided with vehicle-pedestrian parapets, designed to form a barrier for vehicles and to safeguard pedestrians.

Gaps in the grass of the central reserve consisting of a paved area about 20 yards long are provided at intervals of about two miles and at interchanges. These are emergency crossings normally closed to traffic by means of light trellis-type portable barriers. They are intended for emergency use by the police, ambulance and fire services, and for diversion of traffic from a carriageway when it is blocked, or when repairs are being carried out. To enable these emergency crossings to be more easily located by those authorised to use them, a special marker post, two feet six inches high, with two red reflective strips each three inches deep at the top, is provided one hundred yards from the crossing on the verge at each approach.

Blue and white marker posts, also two feet six inches high, are erected normally at intervals of one hundred and ten yards (half-furlongs) on the verge and central reserve. The road faces of the posts on the verges are marked with the symbol of a telephone and an arrow, showing the direction of the nearest emergency telephone. In addition, the marker posts carry numbers indicating the mileage and the number of half-furlongs (for example 220/15) from the zero reference point for the motorway. In the case of motorways radiating from London this zero point is based on the traditional focus point of Charing Cross in London. The marker posts in the verge are predominantly white with a red reflectorised aluminium strip at the top, twelve inches deep by two inches wide. Those in the central reserve are predominantly blue, with a white reflectorised strip.

Emergency telephones are provided normally in pairs at one-mile intervals, replacing the marker posts at these points. The emergency telephone cables now include the required control circuits for a system of illuminated emergency warning signs which are now being installed.

In order to assist in the location of the interchanges, a national system of interchange reference numbers has been evolved, the serial numbers also starting from London, for the motorways radiating from the capital.

The motorway signs have been designed so that they can be easily read by drivers travelling at speed. Where there is considerable traffic movement at an interchange, or the interchange is complex, overhead gantry signs are used to direct traffic into the correct lane well in advance of the interchange itself.

In between the traffic lanes white reflecting studs are provided at a spacing of 54 feet, in the centre of each alternate 18 feet long gap between the nine feet long by four inches wide white line markings. Red studs follow the nearside edge of the deceleration lane at 27 feet spacing and white studs follow the nearside edge of the main carriageway at the same spacing. Between interchanges reflector studs are spaced at 54 feet intervals along the edges of the carriageways, red studs on the nearside and amber studs on the offside. All lane markings and deceleration arrows are reflectorised. These reflecting studs and the reflectorised strips on the marker posts are provided to assist traffic movement at night and in fog conditions and it is important, therefore, that the colours should be widely known.

Travelling along a motorway which has been constructed in stages, the driver will notice that on certain lengths the acceleration and deceleration lanes are parallel to the main carriageway, whereas on other lengths there is a direct taper. On the first motorways in Britain, the acceleration and deceleration lanes were provided with a direct and fairly sharp taper–the Preston By-pass and Lancaster By-pass length of M.6, for example. Subsequently it was decided that these did not provide sufficient length for speed change and later motorways, including the remainder of the M.6 in Lancashire, were constructed with 12 feet wide acceleration and deceleration lanes parallel to the main carriageway. In the light of more recent experience, the current design of the Ministry allows for a longer direct taper, which has been found to provide a better guide to the driver leaving or entering the route. This taper is 1 in 25 on deceleration lanes and 1 in 40 on acceleration lanes, with a radius of about 5,000 feet where the taper meets the main carriageway. For a slip road 20 feet wide, this gives an effective length of 500 feet for a deceleration lane and 800 feet for an acceleration lane, measured to the nose in each case, where the full 20 feet width of slip road is first attained. The double-headed arrow in reflectorised paint on the carriageway is a further guide to the driver in directing him to the required path.

At an exit, the far nose at the point where the full slip road width leaves the main carriageway can be a potential danger, especially at night or under fog conditions. At this point the current design allows for a paved taper to be provided 200 feet long, striped in white, and with red reflecting studs at nine feet centres along both the nearside edge of the main carriageway and the offside edge of the slip road. Similarly, at an entrance on to a motorway, the nose at the start of the acceleration lane is a possible hazard area and here the present design allows for a paved striped taper 300 feet long, again with reflecting studs at nine feet centres, in order to direct traffic leaving the slip road smoothly and tangentially into the main stream of traffic. At the exit and entrance of the slip roads green reflecting studs at 27 feet centres are provided along the nearside edge of the motorway carriageways.

Service areas

On motorways in Britain, service areas are provided initially at intervals of about 24 miles, but it is intended to build additional service areas as the demand increases, so that ultimately the interval will be about 12 miles.

These service areas provide drive-in halts for both commercial and private users with fuel, emergency repair, parking, cafeteria, restaurant and toilet facilities.

At present in England and Wales, there are 16 service areas for the 550 miles of motorway in operation. Each service area is operated by a private developer, who tenders for a 50-year lease, to standards of layout and service prescribed by the Ministry of Transport.

Experience in the operation of earlier service areas and the demands at peak periods has led to the provision of facilities on a larger scale than those originally provided. Evidence of this trend is to be found in the service areas in Lancashire.

The first, opened to use in 1963, was the service area on the M.6 at Charnock Richard, south of Preston, which occupies a total area of about 16 acres, excluding the adjacent motorway maintenance compound (Plate 22). The main feature is the impressive bridge restaurant which spans the M.6 and is particularly attractive at night with its coloured lighting. In addition, there are two transport cafés and two snack bars, and there are rooms for nursing mothers on both sides of the service area. The parking areas provide for about 300 vehicles.

The Forton service area (Plate 23), south of Lancaster on the M.6, which was opened in 1965, occupies an area of about 20 acres and includes a woodland as a picnic area, with its own parking space. The facilities have the spectacular addition of an 80 feet high tower with a restaurant and observation floor, from both of which extensive views over Morecambe Bay and the Lakeland hills can be seen.

For these two service areas, and others already completed, the developer at the time of tendering has submitted his layout design for both the buildings and the parking areas.

The present policy is for service areas to have only limited functions and they provide neither overnight accommodation nor caravan parking for long-distance travellers as it is considered that these can be catered for by existing facilities clear of the motorway. Consideration is being given by the Ministry to the provision of parking areas at scenic vantage points along the motorways in Britain.

Speed limits on motorways

Much has been written–generally critical–about the decision in 1965 of the Minister of Transport to impose, at least temporarily, a speed limit of 70 miles per hour on motorways. I offer no apology for having been a keen advocate of such a speed limit for almost nine years.

However, since I have been so intimately concerned in the designing of motorways, I am conscious that my viewpoint might be coloured by prejudice. I therefore asked 17 senior members of my staff to let me have their views and, at the same time, to list what in their opinion are the advantages and disadvantages of a speed limit. Rather to my surprise they were unanimous in their opinion that the speed limit should be retained.

The following are the principal reasons they gave, both in favour and against the imposition of a speed limit:

Against:

'Under good road and weather conditions many cars are in the hands of experienced drivers which could safely exceed 70 miles per hour. It is extremely frustrating to be late for an appointment with a fast car, well-maintained, and yet be unable to exceed 70 miles per hour.

'Presumably it involves the police in extra work when they may be more beneficially employed elsewhere.

'Many commercial vehicles appear to travel very near to the speed limit and it is obviously unpleasant for the driver of a car to be followed perhaps for many miles by heavy commercial vehicles, travelling possibly too close and at the same speed.

'The imposition of a speed limit means that the driver of a vehicle must keep a continual watch on the speedometer with some distraction from the main requirement of proceeding safely.'

Some replies, although advocating a speed limit, suggested that it should be higher on motorways than on the all-purpose roads.

For:

'The design of rural motorways has been based, from a minimum visibility distance point of view both horizontally and vertically, on 70 miles per hour. Because visibility generally on motorways is so good, drivers may not appreciate that it is limited in isolated cases and in these cases speeds above 70 miles per hour could be dangerous in an emergency.

'There is a curb on recklessness.

'It prevents the motorways being used for high speed trials.

'So long as motorways are unlighted and motorists are driving on dipped headlights, then speeds above 70 miles per hour during times of darkness are not considered safe.

'A speed of 70 miles per hour is quite fast enough for any normal driver, particularly having regard to the narrow central reserve. It is considered common sense to assume that if speeds are restricted the risk of drivers momentarily losing control due to mechanical defect or a burst tyre, will be reduced and, if accidents do occur, the risk of serious or fatal injuries will also be reduced.

'Speed limits can be an effective way of reducing speed if they are recognised by the public as being reasonable. This appears to be the case with the present speed limit, which is more conscientiously observed on motorways than speed limits in general elswhere. A probable reason is that by far the great majority of drivers are content to travel below 70 and this has a chastening effect on the potential limit breakers.

'There is ample evidence from records of use of the emergency telephone system that far too many vehicles are incapable of being run for long periods at high speeds. As traffic increases on the motorways the hazards arising from these causes will grow and with unrestricted speeds would increase even further.

'After travelling some distance on a motorway, there is a tendency for the driver to increase speed without being aware of the fact. Above a certain figure this would put undue strain on the engine and tyres and this would become less likely if the driver was aware of the need to glance at his speedometer from time to time.

'Vehicle design has not advanced sufficiently to provide adequate protection for the occupants in the event of high-speed crashes. There is still no skid-resistant road-surfacing material adequate for high-speed traffic in wet conditions.

'Quite a large proportion of the cars at present manufactured are capable of speeds only slightly in excess of 80 miles per hour. To limit them to 70 miles per hour should keep them within their reasonably safe capabilities.

'One main advantage of the speed limit is that it makes for greater safety when moving into an outer lane to overtake. With a speed limit of 70 the driver can judge better whether it is safe to make the manoeuvre. Some cars are driven at a higher speed than the capability of the driver warrants and a speed limit has some control over this.'

When the speed limit was introduced, it was contended that bunching would occur with resultant multiple vehicle accidents. This has not occurred. On this point the reply of one member of the staff was 'Bunching is bunk'.

'Junctions and interchanges have a safe speed which is usually considerably below that of the motorway. With a speed limit in operation there is less danger of vehicles entering a junction at unsafe speeds.

'With a far greater experience of motorways than we have, every State apart from one in the United States of America has adopted a speed limit.'

Although some drivers will be inconvenienced by the imposition of a speed limit, one must remember that it is only a very small proportion who travel at more than 70 miles per hour, and an even smaller proportion who travel at more than 80 miles per hour. But perhaps the most powerful argument in favour of a speed limit is the experience in Lancashire where, on its 61 miles of M.6, the restriction to a maximum of 70 miles per hour has been followed by a dramatic decrease in the number of casualties; this is shown in Table 7 overleaf, kindly provided by the Chief Constable of Lancashire.

Table 7

Casualties on motorways in Lancashire County Police area

	Casualties			
	Fatal	Serious	Slight	Total
M.6–from 22.12.64 to 21.12.65– No Speed Limit	26	82	220	328
M.6–from 22.12.65 to 21.12.66– 70 m.p.h. Speed Limit	10	53	148	211

For the 12-month period, there was a decrease of 62 per cent in fatal casualties, 35 per cent in serious casualties and 33 per cent in slight casualties, and this is in spite of the fact that the traffic on the motorway in 1966 was 10 per cent more than in 1965.

To me these figures are the final argument in favour of a speed limit.

Chapter 12

Urban motorways

General

The object of this chapter is to highlight the major features of urban as distinct from rural motorways.

Both rural and urban motorways possess the same inherent characteristics of being dual carriageway roads with the opposing streams of traffic separated by a central reserve and provided for the sole use of certain prescribed classes of motor vehicles, fenced off from their surroundings, with complete control of access, grade separation from all other roads and usually with hard shoulders. An urban motorway was defined at the London Conference on Motorways, held in 1956, as simply 'a motorway running through a built-up area'.

In the context of this chapter, however, an urban motorway will be considered to be one for which the standards of layout and design have of necessity to be lower than those for a rural motorway, because of the restrictions imposed by the higher degree of development in built-up areas.

There are two basic types of urban motorways–radial and ring routes. In its idealistic shape, the pattern of urban motorways for a large town is in the shape of a cartwheel with the radial routes forming the spokes of the wheel, and with a circumferential outer ring road similar to the outer rim of the wheel and a circumferential inner ring road similar to the hub of the wheel serving the city centre.

Where possible urban motorways should be sited to follow along existing barriers in development, such as rivers, canals or railways.

Design standards

The standards now referred to are those applicable to urban motorways in Britain.[1] The recommended design speed for urban motorways used as primary distributors

[1] Ministry of Transport, Scottish Development Office, The Welsh Office, *Roads in Urban Areas*, London, Her Majesty's Stationery Office, 1966.

is 50 miles per hour, with a minimum of 40 miles per hour where a higher standard cannot be obtained.

The carriageway width is similar to rural motorways with 12 feet wide traffic lanes and the layout is usually dual two-lane or three-lane. Where possible paved verges on each outer edge of carriageway are provided. These are of a minimum width of five feet where space is tight, but preferably of eight feet in width. Where paved verges are provided they are separated from the carriageway by a marginal strip one foot in width of contrasting colour and texture.

Wherever the five feet width of paved verge cannot be obtained because of severe constriction of space by adjacent development, the nearside edge of each carriageway has a raised kerb and lay-bys are provided at intervals of not more than half-a-mile along each carriageway. The lay-bys are normally ten feet wide and at least 100 feet long plus tapered ends of 53 feet minimum length to allow traffic on the lay-bys to leave or enter the main traffic flow safely.

The standard width of the central reserve on urban motorways is 10 feet, with a minimum width of six feet to incorporate a central safety fence. For narrow central reserves it is usually necessary to have a paved surface instead of grass. Gaps in the central reserve for emergency crossings are required at one-mile intervals, a closer spacing than on rural motorways. Safety fences are required to be at least two feet clear of the nearest edge of carriageway. Bridge piers, retaining walls and lighting columns are required to have at least three feet clearance from the nearest edge of the carriageway.

For a 50 miles per hour design speed the minimum sight distance is 425 feet and the minimum radius should preferably be not less than 1,620 feet with an absolute minimum of 980 feet.

The other cross-section details for urban motorways follow the same principle as for rural motorways.

Interchanges

The subject of interchanges in general is dealt with in Chapter 7 and again the principles of the geometry required for rural motorways apply to urban motorways.

The absolute minimum spacing between interchanges on an urban motorway is usually taken as 1,800 feet, although half-a-mile is preferred. However, great care must be taken to avoid the weaving manoeuvres of traffic entering and leaving the motorway from interfering with the free flow of the through traffic on the motorway.

The basic types of interchanges indicated in Chapter 7 are applicable to urban motorways, but it is possible to adapt various arrangements of slip roads to reduce the spacing between interchanges.

Figure 34 shows a 'scissors' crossing provided to carry one slip road over the other which enables the traffic leaving the motorway for side road B to do so before traffic enters the motorway from side road A.

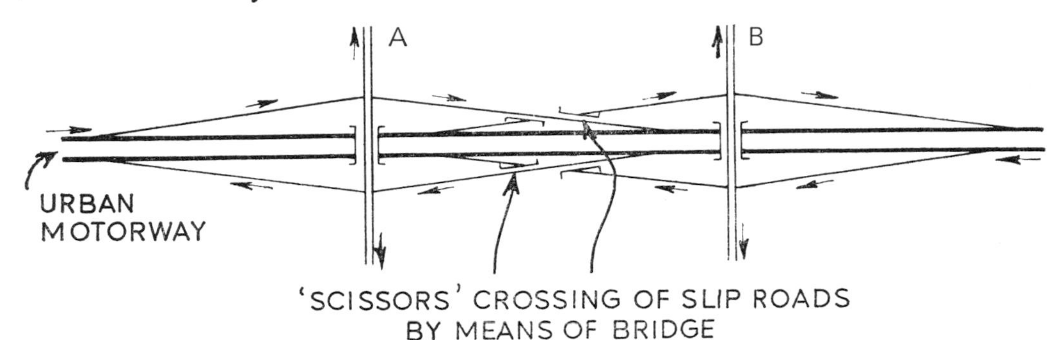

Fig. 34. 'Scissors' crossing of slip roads

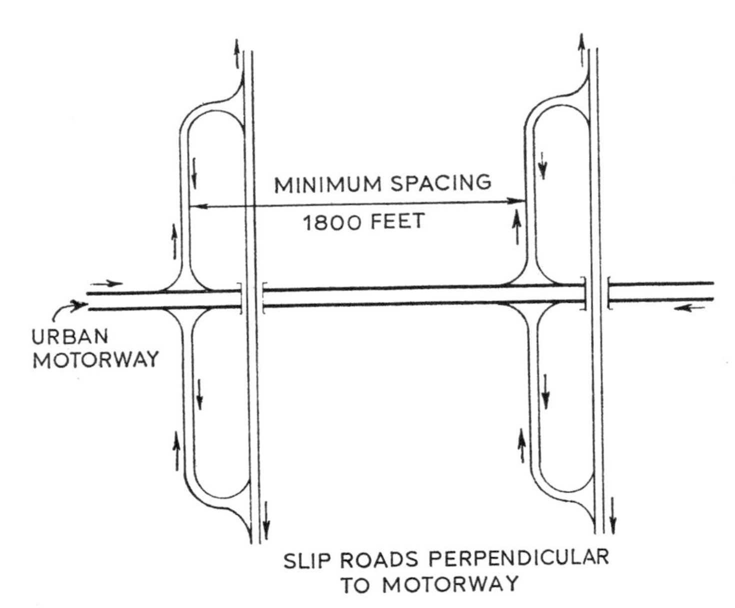

Fig. 35. Slip roads joining motorway at right angles

Where it is possible to locate the slip road at right angles to the motorway, as shown in Figure 35, still providing at least the minimum radius and gradient at the slip road at exit from the motorway, this also enables the interchanges to be spaced closer together where circumstances warrant.

To reduce the number of access points, collector-distributor roads may be provided. Running parallel to the main motorway, and with connections to the all-purpose road system, these roads allow the weaving manoeuvres to take place at a lower speed than on the motorway itself. A diagrammatic layout is shown in Figure 36.

Direct taper acceleration and deceleration lanes are provided for slip roads at interchanges on urban motorways. The lengths of acceleration lanes vary from 800 feet with a 1 in 25 upgrade on the motorway to 440 feet on a 1 in 25 downgrade on the motorway. The comparable lengths for deceleration lanes are 270 feet and 340

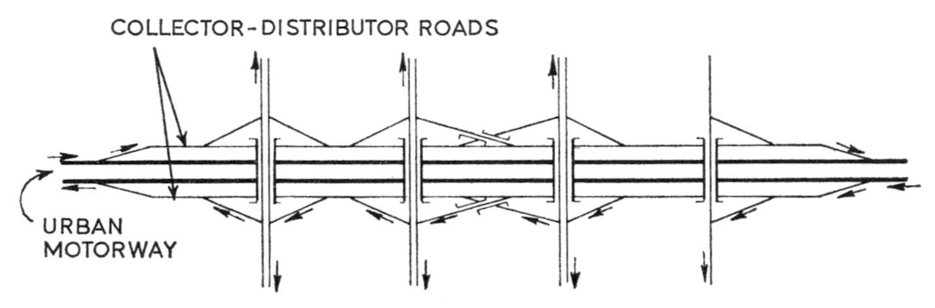

Fig. 36. Collector-distributor roads

feet respectively. Where possible gradients which assist the change in speed are desirable–this means a downgrade for acceleration and an upgrade for deceleration.

Slip roads at interchanges on urban motorways are usually for one-way traffic only and where one lane is sufficient have a width of 14 feet, and for two lanes a width of 24 feet. One lane slip roads have a one-foot wide marginal strip or lip kerb on the nearside, with a paved verge of minimum width of five feet. The minimum design speed for slip roads should be between two-thirds and one-half of the design speed of the urban motorway itself. This gives a minimum radius of about 170 feet for a slip road designed for 25 miles per hour, with a minimum stopping distance of 150 feet. The gradient of slip roads should preferably be not greater than five per cent with an absolute maximum of eight per cent.

As with the design of the motorway itself, the layout of interchanges along an urban motorway should preferably be of a consistent standard with a similar arrangement of exit and entry slip roads so that traffic manoeuvres are simple and can be appreciated easily by drivers.

Tidal and peak hour flow of traffic

When designing any major highway, it is necessary to take into account the tidal flow or directional distribution of traffic (referred to as the D factor) under the design conditions. In other words, when the hourly volume of traffic taken for design occurs, the percentage of traffic travelling in each direction should be known. For example a rural motorway may have 67 per cent of the vehicles travelling on one carriageway at the peak hour. For outer ring roads the figure is generally a little lower. In the central business area for very large cities the D factor could approach a 50 : 50 ratio for both radial and inner ring roads.

One way of allowing for the uneven distribution of peak hour traffic on urban motorways is to provide three carriageways separated by reserves. The central carriageway can be used for traffic into the city during morning peak flow conditions and for traffic out of the city during the evening peak. Links at intervals from the outer carriageways to the central carriageway enable traffic to be switched as required, with removable barriers across the link not in use.

On urban motorways the highest hourly flow is usually in the afternoon peak flow when vehicles are heading for home, rather than the morning peak flow which tends to be spread over a longer period.

Elevated or sunken?

It will be obvious that a motorway which follows closely the profile of the existing ground is invariably cheaper than one which is required to pass over or under obstructions whether they be natural or man-made. However, this is impossible through a built-up area unless all the existing roads in the vicinity run parallel to the motorway. This is very rarely the case and generally the engineer has to decide whether the motorway should be built above or below existing ground level.

Let us first examine the method whereby a cutting is driven through the area and the motorway is constructed at the lower level. In this case the whole of the land occupied by the cutting has to be acquired and any property lying in the way has to be bought and demolished. Bridges to carry roads etc., must be built across the cutting, and services within the route of the motorway need to be diverted. To reduce the land-take involved and the amount of property demolition, the width can be reduced by constructing retaining walls to support the adjoining land instead of forming slopes. These may be very expensive structures and their cost must be considered in relation to the value of land and property saved. Where land values are very high, however, the motorway could be roofed over and the site utilised for car parking, etc.

An alternative is to build the motorway in deep tunnel passing well below all existing development. This is extremely costly and with tunnels of any length lighting and ventilation must be provided, all involving considerable maintenance costs. A further disadvantage is that connections to the motorway cannot easily be provided.

This method, therefore, only lends itself to through routes and cases where there are particular features such as river crossings and local areas of high ground.

An elevated urban motorway can be constructed either with embankments or viaducts. The land occupied by an embankment is sterilised for all time and the completed motorway may provide an objectionable barrier between the severed areas. The viaduct does not have these disadvantages to the same extent but generally costs much more.

From the amenity point of view a sunken urban motorway is more often than not preferable. In an urban area, however, the alteration of existing underground mains and services may be a major factor and in this respect the sunken motorway may prove to be more costly.

A combination of both an elevated and sunken profile may be used.

Structures

The bridges required for an urban motorway are similar in many respects to those needed for a motorway in a rural area. The design loadings are the same and the materials employed do not vary from those normally used.

There are, however, a number of problems which require special consideration. Speed of construction is vital in an urban area in order to minimise the effects on traffic. The design of bridges should, therefore, aim at making as much use as possible of units which can be prefabricated off-site and are easily and speedily erected.

It is probable that the overbridges for side roads will have to carry numerous mains and services including pipes of large diameter. Such bridges may therefore require large pipe bays which may be critical in determining the depth of construction of the deck. In consequence they may have an important bearing on the profile to be adopted, not only for the side roads concerned but also the motorway itself.

The appearance of bridges in an urban area is of major importance. Not only should the design harmonise with the surroundings but the surface treatment to be adopted for exposed faces should be chosen with special care as it must be recognised that they will be seen at close quarters by large sections of the population, especially in the case of concrete surfaces. I have seen notable examples in major cities in other parts of the world where a succession of bridges with untreated concrete surfaces over urban motorways produces an appalling sense of monotony and does nothing to improve the urban environment. If the expenditure of money is justified in making the appearance of buildings, either private or public, more attractive then surely it is equally warranted in the case of bridges built alongside them in our cities and towns. The painting of steel structures in attractive colours is even more important in urban areas and can do much to brighten what may be otherwise drab and uninteresting.

Mention has already been made of the large retaining walls often required in cuttings for urban motorways. These will usually be adjacent to bridges and the design of the two types of structure should be considered together. How best to deal with the large areas of surface involved in order to produce a pleasing appearance at a reasonable cost requires careful consideration. Facing with masonry, concrete blockwork, tiling or brickwork has been used but it is obviously expensive. An alternative, which costs very little more than a plain concrete face, is to apply a finish capable of hiding surface blemishes and imparting a uniform appearance to the concrete. Materials which are essentially paints in character and which can be applied by means of a spray to give a desired texture and colour have been used with success.

Examples of urban motorways

In several of our larger cities in Britain, examples of urban motorways are now in evidence. These are, with a few exceptions, mainly elevated.

Stretford-Eccles By-pass

The Stretford-Eccles By-pass, M.62, when opened to traffic in 1960 was the first motorway in Britain to have the characteristics of an urban motorway.

A scheme for the construction of this six-mile length of motorway, including a high-level bridge over the Manchester Ship Canal, was included in the Ministry of Transport Road Programme in 1953 for commencement in the period 1956–9. In the same year a nearby steelworks was having difficulty in finding an economical tip for its waste slag and, with the approval of the Ministry of Transport, I was able to arrange for this very suitable filling to be tipped and compacted at the site of the largest embankment on the south approach to the high level bridge. The construction of this embankment was carried out by County Council workmen and through using almost 400,000 tons of 'free' material, nearly £100,000 was saved. This embankment was the first physical step taken in the construction of British motorways and was completed at an insignificant cost.

Work on the first contract started in April 1957 only 16 months after the County Council had made the scheme under Section 1 of the Special Roads Act 1949, during which time a public inquiry was held.

An elevated type of motorway, it is carried for $4\frac{3}{4}$-miles of the six-mile length on embankments which rise to a maximum height of 54 feet. Its construction necessitated the importation of over three million tons of filling material.

The number of interchanges indicates the urban character of the motorway. Apart from the terminal junctions, four intermediate connections are provided giving an average spacing of approximately 1·2 miles.

At the northern end of the By-pass–at Worsley–a terminal roundabout was constructed to connect with A.575. A local Preservation Society protested at the Public Inquiry against this proposal because of the detrimental effect they thought it would have on the local surroundings. On completion of the work the County Council received a Civic Trust Award for an outstanding contribution to the appearance of the local scene by virtue of the design and landscaping treatment adopted. This surely indicates that motorways do not necessarily spoil the environment.

Dual 24 feet wide carriageways are provided throughout the length. With the exception of the high-level bridge, the central reserve is 15 feet wide, which includes marginal strips. In accordance with the standards applicable at that time hard shoulders were not provided on the underbridges. Elsewhere, these are eight feet wide and of a similar construction to those originally provided on the Preston and Lancaster By-passes. They are to be paved in the near future.

This first classified road in Britain to have the status of a motorway carries high traffic volumes over its whole length. Of the 30,000 vehicles per day travelling each week day over the high-level bridge, 22 per cent are heavy goods vehicles.

The carriageways are of flexible construction, varying in total depth from 18½ inches to 36 inches. An eight-inch thick limestone 'wet-mix' base was laid on a varying depth of sub-base. This comparatively light specification for the base has stood up remarkably well to the traffic which it has had to carry. The surface is 4½ inches of hot rolled asphalt.

The Barton high-level bridge is described in Chapter 8. The 21 other bridges were constructed with a variety of designs and in view of the urban character of the area, differing forms of surface treatment to the abutments and wing walls were provided.

The northern section of the By-pass is affected by mining subsidence. This is likely to be very severe and it was anticipated that it would occur in varying amounts up to a maximum of 12 feet near the Bridgewater Canal at Worsley. Sufficient headroom has been allowed in the bridges carrying the By-pass over the Canal to provide for this subsidence. Provision has also been made in the design of the bridges for the decks to be jacked up to enable adjustments in level to be made.

The motorway came into use in October 1960.

Highway 401–Toronto

In 1953 a section of Ontario's Trans-Provincial Motorway, known as Highway 401, was opened to traffic across the northern part of the City of Toronto.

At that time it carried 14,000 vehicles per average day, but by 1959 the traffic had increased to 65,000 vehicles per average day. This represented an increase of approximately 29 per cent per year during the six-year period.

The practical capacity of this dual two-lane motorway was considered to be 48,000 vehicles per day at an operating speed of 45 to 50 miles per hour with a maximum possible capacity of 66,000 vehicles per day at an operating speed of 30 miles per hour.

The motorway was therefore reaching a completely congested state with a comparatively high accident record.

As a result of an extensive traffic study carried out in 1959 it was shown that 90 per cent of the traffic volume using the motorway was generated by residents of the city or areas adjacent to it. It was decided that the existing dual two-lane carriageways should be widened to three-lanes for the use of through traffic or traffic travelling relatively long distances.

In addition, a three-lane collector-distributor carriageway was to be provided on either side to serve commuters and local traffic. These were intended to deal with short journeys involving a distance equivalent to three interchanges or less. The entrance and exit transfer lanes connecting the through carriageways to the collector-distributor carriageways were therefore spaced accordingly.

As the distance between interchanges on Highway 401 is comparatively short, the interchanges were designed to connect to the collector-distributor carriageways, leaving the through carriageways comparatively free from weaving movements.

When I visited the city in 1965 the work of reconstruction was in progress. By that time the average daily traffic had risen to 75,000 vehicles per day and exceeded 85,000 vehicles per day at peak periods.

The completed sections indicated clearly the advantages of the collector-distributor system in the design of heavily trafficked urban motorways.

The total cost of the reconstruction of the 19 miles of highway was $63 million.

Postscript

Most of this book has been written during an eventful period of my working life. In 1966 I completed 21 years as County Surveyor and Bridgemaster of Lancashire. Then, on the 1st February 1967, I took up duties on the challenging task of heading the North-Western Road Construction Unit, the first to be set up in this Country.

Until April 1967 the design and supervision of motorways in England were carried out by county councils acting as agents of the Ministry of Transport or by consulting engineers appointed by the Ministry.

A major change has now taken place. The Ministry has divided England into six divisions and motorway work, together with trunk road improvements costing more than £1 million, is entrusted to road construction units based on these divisions. In deciding on the divisions, the Ministry has taken into account the future motorway programme and attempted to base the boundaries on a fairly uniform amount of work for each division.

The object of this new partnership between the Ministry of Transport and the county councils is to expedite the building of motorways and trunk road schemes costing over £1 million, in an endeavour to provide the roads which are essential to our economic survival as a nation.

It is the duty of the Director of each Unit to see that the construction of motorways in the division is carried out economically and to time, and he will decide which work will be given to the county councils within the division to design and which work will be entrusted to consulting engineers. It is the Minister's intention to delegate as many as possible of the functions previously dealt with by the Ministry London Headquarters to each unit.

The units will be staffed jointly by Ministry and county council staff. The units will not deal with overall policy, planning and programming in the wider sense, which remain the responsibility of Ministry Headquarters.

The Minister hopes that the new scheme will have the following advantages:

1. The delegation of many of the Ministry's London functions to the division should help to streamline the procedure and minimise the possibility of delay.
2. The Ministry recently has only been able to recruit a limited number of engineers, as many young engineers have preferred to work on the actual design and construction of motorways. Under the new scheme, Ministry engineers will work with county seconded staff, both in the unit headquarters and in the county sub-units in the division.
3. Under the previous arrangements, a county undertaking the design and supervision of a motorway for the first time often had to advertise for a large number of staff. Under the new arrangements, as the amount of work carried out yearly by the unit should be fairly constant, the number of staff required would also remain constant. Also, staff experienced in major road and bridge construction works should always be available. The scheme, therefore, enables the Minister to draw upon the expertise available to county councils to carry out the technical work of the major national road schemes and will make better use of the manpower available in this field, reducing the competition caused by shortage of engineers when approved schemes move from one county to another.

The building of motorways requires team work. A vast array of people of different skills and abilities are involved in the projects and have their own contribution to make. The range is wide and includes engineers, contractors, lawyers, accountants, surveyors, valuers, town planners, administrators, machine operators, clerks, typists, skilled and unskilled workmen and many others. I would like to take this opportunity to pay tribute to all who help to build motorways. In particular, I wish to place on record my thanks to the staff of the Lancashire County Surveyor's Department, who have worked tirelessly, often for long hours over and above the call of duty, to start, on schedule, the several motorway projects in the county. This successful outcome could not have been achieved without the help and encouragement of the staff of the Ministry of Transport.

I have said that it is team work, but I wish to pay a special tribute to the 'forward line'–the contractors who have the difficult job of translating the designer's ideas into reality, and who, in several instances in Lancashire, have finished the motorways ahead of time.

The building of a motorway is sculpture on an exciting grand scale, carving, moulding, forging and adapting the materials provided by nature–earth, rock and minerals–into a finished product, which must be both functional and pleasing to the eye, as well as economical and durable. But in trying to accomplish this, one must also be humanitarian and remember that all this affects people. The civil engineer on motorway projects, as on other public works, is the servant of the people, using his specialist knowledge on their behalf for the good of the whole community and, at the same time, mindful of the views and rights of the minority who are affected.

I believe profoundly in the value of motorways. They are vital to the national economy and to the smooth flow of exports. As an illustration of this, the two Ports of London and Liverpool handle two-thirds of the country's exports and road transport carries nearly 90 per cent of the export cargo of the Port of London and 80 per cent of that of Liverpool. For passengers too, roads now account for nearly 90 per cent of all internal passenger traffic in this country. With the predicted doubling in the number of private cars in the next ten years, the role of motorways will become even more vital.

I hope that I have been able to convey, through the medium of these pages, some of my enthusiasm for motorways. I have tried to tell part of the story of motorways as I see it, in an easily assimilated way, as free as possible from indigestible statistics and to put forward some of my personal views. It is hoped that this story has been readable and will add to your interest in travelling on motorways.

I wish you happy and safe motoring!

Index

Absorption Factor, 165
Access:
  control of, 21, 27, 30]
  private, 81
Accidents:
  average cost, 73
  fatal, 25
  Germany, 49, 50
  injury, 22, 25
  rate, 22, 25, 73
  saving, 23, 24, 25
  types, 23
Act, Highways 1959, 46, 79
  Special Roads 1949, 36, 45, 46, 205
Agent, Contractor's, 150
Aggregate Crushing Value, 165
Agriculture, Ministry of, 78, 113
Aids to movement of traffic, 192
Air-entraining agents, 162
Alberta, 57
Aldington, Major H. E., 43, 45
Alignment, 97
All-purpose roads, 22, 25
Almondsbury interchange, 125
Alteration to mains and services, 169
Ambulance Service, 187, 191
American Association of State Highway Officials, 59, 125
Antwerp, 54
Aquaplaning, 101
Asphalt, 160
Association of Consulting Engineers, 149
Austria, 54
Autobahnen, 17, 29, 30, 37, 42, 50, 52
Autoroutes:
  du Sud, 51
  Belgium, 52
Autostrade:
  Milan-Varese, 27
  Milan-Turin, 27
  Naples-Pompeii, 27
  del Sole, 50
  Bologna-Bari, 50
  Naples-Bari, 50
A.6, 24, 25

Baker, J. F. A., 46
Baker, Noel, 40
Bankettes, 180
Bari, 50
Barnes, Alfred, 43
Barton high level bridge, 45, 145, 206
Base, 108, 110, 160, 172, 175, 179
Belgium, 51, 52, 54
Benefits, 26
Bill of Quantities, 84, 154, 166, 170
Birmingham, 35, 39, 40, 46
Bitumen:
  macadam, 160
  petroleum, 161
  pitch, 161
Blackpool Ring Road, 17
Bologna, 50
Boyd-Carpenter, J. A., 46
Brenner, 54
Bridges:
  appearance, 143
  bearings, 141
  clearances, 136
  construction, 171
  design loading, 135
  expansion joints, 141
  foundations, 140
  maintenance, 186
  materials, 137
  parapets, 142
  piles, 141
  principal features, 135
  spans, 137
  surfacing, 142
  types, 138
Bridgewater Canal, 206
Bristol-Birmingham Motorway M.5, 45, 46, 125
Britain, 34, 43
British:
  Columbia, 57
  Railways, 78
  Road Federation, 43, 47, 131
  Road Tar Association, 161
  Standards Institution, 154
  Waterways, 78
Bronx River Parkway, 30
Broughton interchange, 126
Brussels, 52
Bureau of Public Roads, 57, 59, 60
Burgin, Leslie, 37

California, 31, 109, 125
California Bearing Ratio, 109
Canada, 56
Capetown, 54
Carriageway:
  capacity, 25, 73, 97
  width, 100, 101
  works, 172
Castle, Mrs. Barbara, 49
Cement and Concrete Association, 30
Central reserve, 101

Centrifugal force, 99, 100
Certificates and payment, 153
Charlesworth, Dr. G., 71
Charnock Richard Service Area, 195
Chief Constable, 197
Civic Trust Award, 206
Clay, 107, 157
Cohesive Soils, 108
Collector-distributor roads, 122, 202, 207
Common Market, 51
Compaction, earthworks, 158
Compensation, 92, 94
Compulsory Purchase Order, 82, 92, 93
Connecticut, 62
Concrete:
 for structures, 164
 lean, 108, 160
 pavement, 108, 110, 111, 161, 180
 pre-stressed, 164
Construction, 169
Contract:
 Conditions of, 149, 166
 Drawings, 84, 86, 154
 Procedure, 149, 167
Contractor, 87, 95, 149, 180, 209
Cook, Sir Frederick, 40
Cook, R. Gresham, 43
Copenhagen, 53
County:
 Boroughs, 78
 Councils, 78
 Councils' Association, 37
 Motorways, 95
 Surveyor, 71, 78, 150
 Surveyor's Department, 209
 Surveyors' Society, 17, 36' 40, 43
Croft interchange, 133
Crossfall, 100
Cross-section, 101
Culverts, 114, 170
Curvature, 98

Dawson, R. F. F., 71
Denmark, 53, 54
Design:
 Charts, 109
 motorway, 96
 speed, 49, 64, 99, 100, 115, 116, 121, 122, 200, 202
Direct Labour, 176
District Valuer, 82, 88, 91
Divided highway, 31
Drainage:
 carriageway, 99
 construction, 172
 design, 112
 specification, 155
Drawings, 84, 86, 154
Durban, 55

Earthworks, 156, 170, 171
East Lancashire Trunk Road A.580, 72, 128, 129, 130
Eccles interchange, 132
Economic assessments, 75
Economics of Motorways, 43
Efficiency in Road Construction, 68, 151, 167
Elevated urban motorway, 203
Employer, 149
Engineer, to the contract, 149
Europe, 42, 54, 55
Excavation, 156
Expansion joints:
 bridges, 141
 pavements, 111, 112, 181
Expressway, 32
E. 3, 52

Federal Aid, 57, 59, 66
Federation of Civil Engineering Contractors, 149
Fences:
 permanent, 155
 temporary, 95
Fire Brigade, 187, 190
Flakiness Index, 165
Flexible Pavement, 108, 109, 160
Fog, 185, 190, 191
Footpaths, 81, 135
Formation, 158, 159, 173
Formwork, 163
Forton Service Area, 116, 195
Foundations, 107, 108, 140
France, 51, 52
Fraser, Tom, 182
Freeway, 32, 62
French drains, 102, 113, 114, 156
Frost:
 damage, 175
 penetration, 110
Fylde Junction:
 —higher bridge, 127, 128
 —lower bridge, 127, 128

Gathurst Viaduct, 147, 177, 179
Generated traffic, 76
Geological Survey, 105
German:
 Roads Delegation, 36
 State Motor Roads, 28, 38
 State Railways, 28
Germany, visit 1938, 17, 36
 mileage, 50
 post-war, 49
Ginns, H. N., 47
G.P.O., 45
Grade separated interchange, 115
Gritting, 184
Guard rails, 155, 186, 192

Hamilton, 56
Hard shoulders, 43, 49, 101, 161, 176, 186
Harris, Sir William, 49
Highway 401, Toronto, 56, 206
Highway Trust Fund, 59
Highways Act 1959, 46, 79
Hill, Dr. Charles, 46
History, 27
Holland, 52, 53
House of Lords, 37, 38, 39
 Commons, 38, 43
Housing and Local Government, Ministry of, 81

Ice warning, 190
Imported material, 157
Inquiry, Public, 81, 93
Inspector, 81, 93
Institution of:
 Automobile Engineers, 43
 Civil Engineers, 39, 43, 47, 149, 150
 Electrical Engineers, 43
 Highway Engineers, 36, 37, 40, 43
 Mechanical Engineers, 43
 Municipal and County Engineers, 40, 43
Insurance, 152
Interchanges:
 design, 115
 design procedure, 125
 diamond, 119
 examples, 126
 four-level, 124, 133
 free-flow terminal, 121
 full cloverleaf, 121
 grade separated roundabout, 119, 122
 partial cloverleaf, 120
 spacing, 116
 split diamond, 119
 three-level, 122, 133
 trumpet, 121, 180
 two-level, 117

types, 117
urban motorway, 200
Interstate System, 57, 58, 59, 60, 66
Italy, 27, 50, 54

Japan, 56
Johannesburg, 55
Joints:
bridge expansion, 141
concrete pavement, 111, 112, 181
Journey time, reduction, 25, 72, 73, 76
Justification, 70

Lancashire, 17, 22, 26, 53, 115, 173, 175, 179, 183, 208
Lancashire County Council, 18, 37, 45, 93, 176, 183, 205
Lancashire-Yorkshire Motorway, 45, 72, 79, 96, 97, 120, 122, 128, 130, 131, 132, 134
Lancaster By-pass, 25, 46, 126, 147, 148, 180, 187, 194
Landscaping:
Ministry's Advisory Committee, 79, 102
Land:
acquisition, 88
reference plans, 89, 90
reference schedules, 89, 90
Section, 88
Lands Tribunal, 93
Lay-bys, 29, 43
Le Bourget, 51
Lennox-Boyd, A. T., 17
Lille, 52
Line:
fixing the, 79
white, 173, 186
Lisbon, 52
Liverpool, 26, 35, 210,
outer ring road, 133, 134
Lloyd, Lord, 45
London, 35, 39, 40, 210
—Birmingham Motorway M1, 46, 131
Long Island, 31
Los Angeles, 125
Lune Bridge, 147
Lyddon, A. J., 43
Lyons, 51, 52

Macadam, John Loudon, 112
Macmillan, Harold, 46, 173
Mains and services, 169
Maintenance, 77, 153, 183
Manchester, 35
—Preston Motorway M.61, 71, 128, 129, 130
Ship Canal, 45, 145, 146, 205
Marples, Ernest, 46
Marginal strips, 101, 172
Marseilles, 51, 52
Materials and testing, 165
Medway Towns Motorway, 46
Mersey, 26, 147
Method of Measurement, 166
Milan, 27, 50
Mileage, 47
Mine workings, 107, 178
Montagu, Lord, of Beaulieu, 34, 35
Montgomery, 27
Montreal, 57
Moses, Robert, 31, 32
Munich, 54
Mussolini, 27
M.1, 46, 131
M.4, 125
M.5, 45, 46, 125
M.6, 18, 21, 24, 25, 26, 38, 46, 92, 112, 116, 119, 121, 126, 127, 133, 144, 146, 149, 171, 173, 176, 179, 180, 182, 183, 184, 185, 187, 188, 189, 191, 194, 195, 197, 198
M.50, 45, 46
M.61, 71, 129, 130
M.62, 25, 45, 46, 72, 96, 116, 122, 128, 129, 130, 131, 132, 133, 134, 144, 145, 205
M.62/M.6 Interchange, 132
M.62/Liverpool Outer Ring Road Interchange, 133

Naples, 27, 50
National Coal Board, 78, 105, 178
Farmers' Union, 78, 79, 93
Trust, 79
National Roads Board:
Italy, 51
South Africa, 55
Network Analysis, 69
New Jersey Turnpike, 62, 63, 64, 65, 66
New Towns, 72
New York City, 30, 32, 35
Nice, 52
North-East Lancashire Motorway, 97, 122
Norway, 53

Objections, 81, 82, 93, 94
Order:
Compulsory Purchase, 82, 92, 93
Section 7, 82
Section 9, 82
Section 13, 82
Ontario, 56, 206
Ordnance Survey, 105
Ostend, 52
Overloading, 73

Paisley, J. L., 71
Parapets, 142
Paris, 51
Parkway, 32
Bronx River, 30, 31
Passenger car unit (p.c.u.), 97
Pavement, type of, 108
Peat, 131
Peak-hour flow, 97, 202
Pedal cyclists, 22
Pedestrians, 22
Pennsylvania, 33
Turnpike, 33, 34
Piezometers, 177
Pipes, 155
Plans:
plot, 82
reference, 82
Planning:
Acts, 157
Department, 78, 84
Plant, earthmoving, 171
Planting, shrubs, 103
Plastic limit, 157
Polders, 53
Police, 22, 78, 187, 188
Polished Stone Value, 165
Pompeii, 27
Press, 88
Preston By-pass M.6, 17, 21, 25, 45, 46, 72, 110, 126, 127, 173, 179, 180, 187, 188, 194
Preston-Lancaster Motorway M.6, 18, 112, 116, 171, 179
Preston Northerly By-pass, 75, 77
Pre-stressed concrete, 137, 139, 164
Pretoria, 55
Procedure, 67
Programme, Ministry's road, 43, 44, 46, 70, 71, 205
contractor's, 151
Protection against atmospheric corrosion, 165
Public Inquiry, 81, 93
Purchase Notice, 94
Puricelli, P., 27
Pyrenees, 54

INDEX

Quantities, Bill of, 84, 154, 166, 170
Quebec, 56, 57
Queen Elizabeth Way, 56

Radius of curvature, 98, 99, 103
Rate of return, 71
Reflecting roadstuds, 101, 186, 193
Regional Land Commissioners, 81
Re-housing, 95
Reichsautobahnen, 28
Resident Engineer, 18, 86, 150
Reynolds, D. J., 71
Rigid pavement, 108, 110, 111
River Authorities, 78, 113
Road:
  Board, 34
  Construction Unit, 49, 208
  Improvement Association, 34
  Note 29, 109, 110, 159
  Plan for Lancashire 1949, 23
  Research Laboratory, 24, 30, 70, 71, 113
Roads Campaign Council, 47
Rock, 106, 157
Rommel, 27
Ross Spur Motorway, 45, 46
Royal Fine Art Commission, 84
Runcorn New Town, 75
Rural motorway, 97

Safer travelling, 24, 73, 76
Safety fences, 155, 192
Salford, 75
Salzburg, 54
Samlesbury bridge, 176
Sand drains, 177
Scale, 104, 105, 154
Scheme, Section 11, 79
Schlamme, 101, 161, 182
Schofield, Peter, 17
Scissors crossing, 201
Scott, Sir Leslie, 36
Section 1, 205
Section 7, 82
Section 9, 82
Section 11, 79
Section 13, 82
Selecting the route, 78
Service Areas, 194
  Charnock Richard, 195
  Forton, 116, 195
Settlement gauges, 177
Setting-out, 151
Severn Bridge, 46
Shale, burnt red, 175
Side roads, 81
Signs, 185, 193
Sight distance, 97, 98
Site clearance, 155
Slip circle, 107, 108
Slip roads, 104, 120, 122, 194, 201
Slurry seal, 161
Snow clearing, 184
Snowhill Lane bridge, 85
Society of Motor Manufacturers and Traders, 43
Soil:
  strip, 170
  Survey, 105, 156, 157
South Africa, 54
South Lancashire Motorway, 75, 131, 132, 133, 134
Special Roads Act 1949, 36, 45, 46, 205
Specification, 154
Speed Limits, 195
Standard:
  British, 154
  design, 96
  Method of Measurement, 166
State Highway Departments, 59
State Holding Company, 51
Statutory Undertakings, 78, 169
Steelwork:
  riveted, 165
  welded, 165
Stockholm, 52
Stretford-Eccles By-pass M.62, 45, 95, 116, 131, 144, 145, 205
Structures, 163
Sub-base, 110, 159, 172
Sub-contractor, 151
Sub-grade, 110
Suitable material, 156
Sunken urban motorway, 203
Superelevation, 99
Surface regularity, 163
Surfacing, 160, 172
  skid resistance of, 22, 163, 181
Survey:
  aerial, 105
  ground, 105
  origin and destination, 72, 73
  route, 104
  soil, 105
Sweden, 53
Switzerland, 54
Tarbock interchange, 134
Tarmacadam, 160
Telephones, emergency, 49, 101, 187, 189, 193
Temporary:
  surfacing, 175
  works, 152
Testing, 165
Texture depth, 162
Thelwall:
  bridge, 146, 183
  to Preston, 176
Tidal flow, 31, 202
Todt, Dr. Fritz, 29
Tokyo, 56
Toll road, 32, 51, 61, 62, 63, 64, 65, 66
Toronto, 56, 206
Traffic, 96
Traffic Commissioners, 81
Trafford Park Industrial Estate, 117
Transition curve, 100
Transport:
  Institute of, 43
  Minister of, 17, 37, 38, 43, 45, 46, 68, 70, 78, 81, 82, 87, 93, 126, 179, 182
  Ministry of, 18, 21, 37, 40, 43, 46, 67, 68, 70, 71, 73, 79, 81, 82, 92, 93, 95, 96, 97, 98, 99, 101, 105, 109, 112, 116, 157, 173, 183, 184, 185, 186, 195, 205, 208, 209
Trans-Canada Highway, 57
Trees, 104, 185
Tripoli, 27
Turin, 51
Turnpike, 32, 33, 34, 61, 62, 63, 64, 65, 66

Unsuitable material, 156
Urban motorways, 18, 50, 199, 205
U.S.A.:
  Alaska, 61
  Bureau of Public Roads, 57
  construction of express highways, 28
  development in the, 30
  District of Columbia, 61
  National System of Interstate and Defense Highways, 57, 58, 66
  Speed Limits, 197
  Toll Roads, 61
  Vehicle ownership, 42
  visits to, 18, 34, 112

Vancouver, 57
Varese, 27
Variation of Price Clauses, 166
Variations, 153
Vienna, 54

Warrington, 76
Watkinson, H. A., 46
West Germany, 49, 50
Wet-mix, 175, 177
Wolverhampton, 35
Worcestershire, 45
Workmanship, 165
Worsley Braided interchange, 128

Yorkshire Road, 72
Yugoslavia, 54